Virtual Machining
Using CAMWorks 2023

Kuang-Hua Chang, Ph.D.
School of Aerospace and Mechanical Engineering
The University of Oklahoma
Norman, OK

SDC
PUBLICATIONS

SDC Publications
P.O. Box 1334
Mission, KS 66222
913-262-2664
www.SDCpublications.com
Publisher: Stephen Schroff

ISBN-13: 978-1-63057-572-4
ISBN-10: 1-63057-572-0

Printed and bound in the United States of America.

Preface

Virtual machining is the use of simulation-based technology, in particular, computer-aided manufacturing (CAM) software, to aid engineers in defining, simulating, and visualizing machining operations for parts or assembly in a computer, or virtual, environment. By using virtual machining, the machining process can be defined and verified early in the product design stage. Some, if not all, of the less desirable design features in the context of part manufacturing, such as deep pockets, holes or fillets of different sizes, or cutting on multiple sides, can be addressed while the product design is still being finalized. In addition, machining-related problems, such as undesirable surface finish, surface gouging, and tool or tool holder colliding with stock or fixtures, can be identified and eliminated before mounting a stock material on a CNC (computer numerical control) machine at the shop floor. In practice, the toolpaths generated can be converted into NC code (also called G-code or M-code) to machine functional parts as well as die or mold for part production. In most cases, the toolpath is generated in a so-called CL (cutter location) data format and then converted to G-code using respective post processors. Moreover, manufacturing cost, which constitutes a significant portion of the product cost, can be estimated using the machining time estimated in the virtual machining simulation.

In this book, *Virtual Machining Using CAMWorks 2023*, we discuss the concept and steps in conducting virtual machining using CAMWorks. CAMWorks, offered by HCL Technologies Limited (www.camworks.com), is a parametric and feature-based virtual machining software embedded in SOLIDWORKS as a fully integrated module. In fact, CAMWorks is the first SOLIDWORKS certified gold product for CAM software, which provides excellent capabilities for support of machining simulations created directly on solid models of SOLIDWORKS. The book covers basic concepts and frequently used commands and options required for readers to advance from a novice to an intermediate level in using CAMWorks. Basic concepts and commands introduced include extracting machinable features (such as 2.5 axis features), selecting machine and cutting tools, defining machining parameters (such as feedrate, spindle speed, depth of cut, and so on), generating and simulating toolpath, and post processing CL data to output G-code for support of physical machining. The concept and commands are introduced in a tutorial style presentation using simple yet realistic examples. Both milling and turning operations are included. Example files are prepared for readers to go over all lessons in this book. Readers may download them from the publisher's website: www.sdcpublications.com.

One unique feature of this book is the incorporation of the CL data verification by reviewing the G-code converted from the toolpaths. This helps readers understand how the G-code is generated by using the respective post processors, which is an important step and an ultimate way to confirm that the toolpaths and G-code generated are accurate and useful.

This book is intentionally kept simple. It primarily serves the purpose of helping readers become familiar with CAMWorks in conducting virtual machining for practical applications. This is not a reference manual of CAMWorks. You may not find everything you need in this book for learning CAMWorks. But this book provides you with basic concepts and steps in using the software, as well as discussions on the G-code generated. After going over this book, readers should gain a clear understanding in using CAMWorks for conducting virtual machining simulations and should be able to apply the knowledge and skills acquired to carry out machining assignments and bring machining consideration into product design.

This book should serve well for self-learners. A self-learner should have basic physics and mathematics background, preferably a bachelor's or associate degree in science or engineering. We assume that readers are familiar with basic manufacturing processes, especially milling and turning. Details related to basic milling, turning, and hole making can be found in excellent textbooks, such as *Manufacturing, Engineering & Technology,* 6th ed., by Serope Kalpakjian and Steven R. Schmid. Familiarity with G-code is extremely important in learning virtual machining and transitioning from virtual to physical machining. Therefore, we encourage readers to review NC programming books, for example, *Technology of Machine Tools,* 7th ed., by Krar, Gill, and Smid. If you are interested in understanding how the toolpaths are generated, i.e., the theory and behind-the-scenes computation, you may refer to books, such as *e-Design, Computer-Aided Engineering Design*, or *Product Manufacturing and Cost Estimating using CAE/CAE*, written by the author. And certainly, we expect that readers are familiar with SOLIDWORKS part and assembly modes. A self-learner should be able to complete the ten lessons of this book in about forty hours. An investment of forty hours will advance readers from a novice to an intermediate user level, a wise investment.

This book also serves well for class instruction. Most likely, it will be used as a supplemental reference for courses like CNC Machining, Design and Manufacturing, Computer-Aided Manufacturing, or Computer-Integrated Manufacturing. This book should cover five to six weeks of class instruction, depending on the course arrangement and the technical background of the students. Some of the exercise problems provided at the end of individual lessons may take noticeable effort for students to complete. The author strongly encourages instructors or teaching assistants to go through those exercises before assigning them to students.

For those who desire to learn more about CAMWorks, you may find additional references on the computer where CAMWorks is installed. Several tutorial manuals provided by Geometric can be found on your computer where CAMWorks is installed:

C:\Program Files\CAMWorks2023x64\CAMWorks_VC142\Lang\English\Manuals

with example files located at:

C:\CAMWorksData\CAMWorks2023x64\Examples

Also, a few useful videos can be found on YouTube (reviewed in June 2023):

www.youtube.com/watch?v=JLt9HNvfjmA (general introduction)
www.youtube.com/watch?v=QnDBDUXdXa0&feature=related (2.5 axis milling)
www.youtube.com/watch?v=XuzBEHXy2iQ (3 axis milling)

and websites of technical consulting firms, such as:

Hawk Ridge Systems: www.hawkridgesys.com/products/camworks
GoEngineer: www.goengineer.com/solidworks/cam

Happy CAMWorking!!

KHC
Norman, Oklahoma
June 10, 2023

Acknowledgements

My sincere appreciation is due to Mr. Stephen Schroff at SDC Publications for his encouragement and support for converting the book idea into reality. Without his support and the help of his staff at SDC Publications, in particular Mr. Zach Werner, this book would still be in its primitive stage.

Thanks are due to undergraduate students at the University of Oklahoma (OU) for their help in testing the examples included in this book. They made numerous suggestions that improved clarity of presentation and found numerous errors that would have otherwise crept into the book. Their contributions are greatly appreciated.

I am grateful to my former students, Dr. Yunxiang Wang and Mr. Peter Staub, for their excellent efforts in carrying out the sheet metal forming project that produced results included in Lesson 10. The technical support provided by engineers and shop floor technicians, including Chris Montalbano, Mark Lucash, Jason Mann, Todd Bayles, Nate Pitcovich, and David Mason, who contributed to the success of the project is acknowledged and is highly appreciated.

About the Author

Dr. Kuang-Hua Chang is a professor for the School of Aerospace and Mechanical Engineering at the University of Oklahoma (OU), Norman, OK. He received his diploma in Mechanical Engineering from the National Taipei Institute of Technology, Taiwan, in 1980; and M.S. and Ph.D. degrees in Mechanical Engineering from the University of Iowa in 1987 and 1990, respectively. Since then, he joined the Center for Computer-Aided Design (CCAD) at Iowa as a Research Scientist and shortly after was promoted as CAE Technical Area Manager. In 1997, he joined OU. He teaches mechanical design and manufacturing, in addition to conducting research in computer-aided modeling and simulation for design and manufacturing of mechanical systems.

His work has been published in 10 books and more than 150 articles in international journals and conference proceedings. He has also served as technical consultant to US industry and foreign companies, including LG-Electronics, Seagate Technology, etc. He served as Associate Editor for two international journals: *Mechanics Based Design of Structures and Machines* and *Computer-Aided Design and Applications*.

About the Cover Page

The picture on the book cover was captured from a computer screen showing a die and a solid model of a part, which is a half clamp of a fuel line in an aerospace engineering system. The die is part of the tooling, which consists of the die, a punch, and blank, employed to carry out sheet metal forming for manufacturing the half clamp. In this application, the CAD model of the half clamp was created in SOLIDWORKS. The die face was designed based on the part geometry and the punch was created by offsetting the die face with the part thickness. Thereafter, sheet metal forming simulations were carried out using DynaForm (www.eta.com/inventium/dynaform) that helped explore the formability of the part and identify a set of feasible process parameters. Such simulations suggest a narrow window of forming process that the part would be successfully formed without tearing and excessive wrinkles. The finalized die surface geometry was exported from DynaForm and imported into SOLIDWORKS for tooling design. Toolpaths were generated for the die and punch machining operations using CAMWorks. A virtual HAAS mill was added to CAMWorks for visualizing machine simulations, in which collisions were detected between tool holder and the stock material in the final finish operation when the tool was reaching the bottom of the pockets of the die. These issues were addressed by selecting a longer tool and a smaller tool holder. The remedy was soon implemented on the shop floor and the collision issues were eliminated successfully. Die and punch were physically machined using the G-code post processed from the toolpaths generated in CAMWorks. The machined tooling met the functional and quality requirement for part production. The half clamp was formed successfully on the shop floor using a 300-ton four-post press. This was a successful industrial application, where CAMWorks played a critical role in creating toolpaths to manufacture the tooling for support of sheet metal forming. More about this application and the use of CAMWorks can be found in Lesson 10: Die Machining Application.

Table of Contents

Lesson 1: Introduction to CAMWorks

1.1 Overview of the Lesson

CAMWorks, offered by HCL Technologies Limited (www.camworks.com), is a parametric, feature-based virtual machining software. By defining areas to be removed as machinable features, CAMWorks is able to apply more automation and intelligence into CNC (Computer Numerical Control) toolpath creation. This approach is more intuitive and follows the feature-based modeling concepts of computer-aided design (CAD) systems. Consequently, CAMWorks is fully integrated with CAD systems, such as SOLIDWORKS (and Solid Edge and CAMWorks Solids). Because of this integration, you can use the same user interface and solid models for product design and later to create machining simulations. Such a tight integration completely eliminates file transfers using less-desirable standard file formats such as IGES, STEP, SAT, or Parasolid. Hence, the toolpaths generated are on the SOLIDWORKS part, not on an imported approximation. In addition, the toolpaths generated are associative with SOLIDWORKS parametric solid model. This means that if the solid model is changed, the toolpaths are changed automatically with no or minimal user intervention. In addition, CAMWorks is available as a standalone CAD/CAM package, with embedded CAMWorks Solids as an integrated solid modeler.

One unique feature of CAMWorks is the AFR (automatic feature recognition) technology. AFR automatically recognizes over 20 types of machinable features in solid models of native format or neutral file format, including mill features such as holes, slots, pockets and bosses; turn features such as outside and inside diameter profiles, faces, grooves and cutoffs; and wire EDM features such as die openings. This capability is complemented by interactive feature recognition (IFR) for recognizing complex multi-surface features, as well as creating contain and avoid areas.

Another powerful capability found in CAMWorks is its technology database, called TechDB™, which provides the ability to store machining strategies feature-by-feature, and then reuse these strategies to facilitate the toolpath generation. Furthermore, the TechDB™ is a self-populating database which contains information about the cutting tools and the machining parameters used by the operator. It also maintains information regarding the cutting tools available at the shop floor. This database within CAMWorks can be customized easily to meet the user's and the shop floor's requirements. This database helps in storing the best practices at a centralized location in support of machining operations, both in computers and at the shop floors.

We set off to learn virtual machining and explore capabilities offered by CAMWorks in this lesson. The follow-up lessons and examples offered in this book are carefully designed and structured to support readers becoming efficient in using CAMWorks and competent in carrying out virtual machining simulation for general applications.

We assume that readers are familiar with part and assembly modeling capabilities in SOLIDWORKS, comfortable with NC programming and G-code, and understand the practical aspects of setting up and conducting machining operations on CNC machines at the shop floor. Therefore, this book focuses solely on illustrating virtual machining simulation using CAMWorks for toolpath and G-code generations. Topics, such as NC part programming and transition from virtual machining to practical NC operations, can be referenced in other books mentioned in the preface of the book.

1.2   Virtual Machining

Virtual machining is a simulation-based technology that supports engineers in defining, simulating, and visualizing machining processes in a computer environment using computer-aided manufacturing (CAM) tools, such as CAMWorks. Working in a virtual environment offers advantages of ease in making adjustment, detecting error and correcting mistakes, and understanding machining operations through visualization of machining simulations. Once finalized, the toolpath can be converted to G-code and then uploaded to a CNC machine at the shop floor to physically machine parts.

The overall process of using CAMWorks for conducting virtual machining consists of several steps: create design model (solid models in SOLIDWORKS part or assembly), choose NC machine and create stock material (or stock for short), extract or identify machinable features, generate operation plan, generate toolpath, simulate toolpath, and convert toolpath to G-code through a post processor, as illustrated in Figure 1.1. Note that before extracting machinable features, you must select an NC machine, i.e., mill, lathe, or mill-turn, choosing tool cribs, selecting a suitable post processor, and then creating a stock material

The operation plan involves the NC operations to be performed on the stock, including selection of part setup origin, where G-code program zero is located. Also included is choosing tools, defining machining parameters, such as feedrate, stepover, depth of cut, etc.

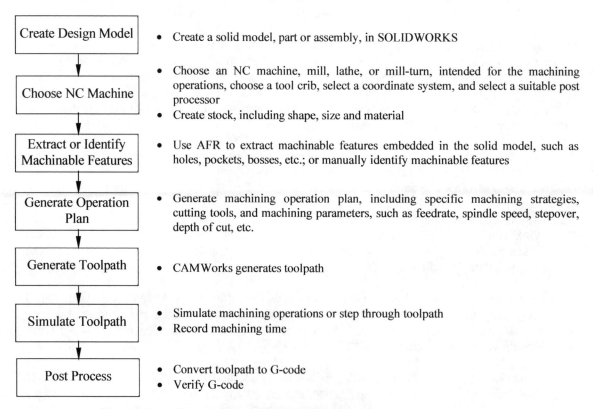

Figure 1.1 Process of conducting virtual machining using CAMWorks

Note that operation plans are automatically determined by the technology database of CAMWorks as long as a machinable feature is extracted automatically or identified manually. Users may make changes to any part of the operation plan, for instance, choosing a different tool (also called cutter, or cutting tool in the book), entering a different feedrate, adjusting depth of cut, etc. After an operation plan is defined, CAMWorks generates toolpath automatically. Users may simulate material removal process, step through machining toolpath, and review important machining operation information, such as machining time that contributes partially to the product cost.

The design model (also called part, design part or target part in this book), which is a SOLIDWORKS part representing the perfectly finished product, is used as the basis for all machining operations. Machinable features are extracted automatically or identified interactively on the design model as references for individual toolpaths. By referencing the geometry of the design model, an associative link between the design model and the stock is established. Because of this link, when the design model is changed, all associated machining operations are updated to reflect the change.

The following example, a block with a pocket and eight holes shown in Figure 1.2, illustrates the concept of conducting virtual machining using CAMWorks. The design model consists of a base block (a boss extrude solid feature) with a pocket and eight holes that can be machined from a stock of rectangular block (the raw stock shown in Figure 1.3) through pocket milling and hole drilling operations, respectively. A generic NC machine *Mill-in.* (3-axis mill of inch system) available in CAMWorks is chosen to carry out the machining operations. For example, toolpaths for machining the pocket (both rough and contour milling operations), as shown in Figure 1.4, can be generated referring to the part setup origin located at the top left corner of the stock (see Figure 1.3).

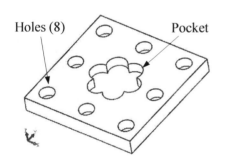
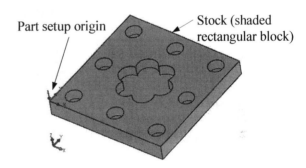

Figure 1.2 Design model in SOLIDWORKS   Figure 1.3 Stock enclosing the design model

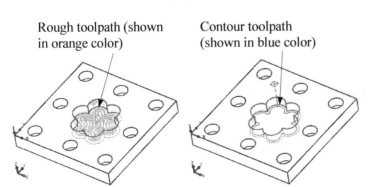

Figure 1.4 Toolpath of the pocket milling operations

Figure 1.5 Step through toolpath

Users can step through the toolpaths, for example, the contour milling operation for cutting the pocket with tool holder turned on for display, as shown in Figure 1.5. The material removal simulation of the same toolpath can also be carried out like that of Figure 1.6.

1.3 CAMWorks Machining Modules

The machining modules included in CAMWorks represent a fairly complete set of capabilities in support of virtual machining and toolpath generations. These modules include:

- 2½ axis mill: includes roughing, finishing, thread milling, face milling and single point cycles (drilling, boring, reaming, tapping) to machine prismatic features;
- 3 axis mill: includes 2.5 axis capabilities plus strategies to machine complex, contoured surfaces encountered in mold making and aerospace applications;
- 2 and 4 axis turning: includes roughing, finishing, grooving, threading, cutoff and single point cycles (drilling, boring, reaming and tapping);

Figure 1.6 The material removal simulation

- Mill-turn: includes milling and turning capabilities for multitasking machine centers;
- Multiaxis machining: 4 axis and 5 axis machining, including high-performance automotive part finishing, impellers, turbine blades, cutting tools, 5 axis trimming, and undercut machining in mold and die making;
- Wire EDM: 2.5 axis and 4 axis cutting operations automate the creation of rough, skim and tab cuts.

All the above capabilities, except wire EDM, are discussed in this book. In addition, CAMWorks supports machining of multiple parts in a single setup. Parts are assembled as SOLIDWORKS assembly, which includes parts, stock, clamps, fixtures, and jig table in a virtual environment that accurately represent a physical machine setup at shop floor. A multipart machining example, as shown in Figure 1.7, with ten identical parts in an assembly will be introduced in Lesson 5. Furthermore, machining features on multiple planes of parts mounted on the respective four faces of the tombstone, as shown in Figure 1.8, in a single setup is supported. More about multiplane machining operations can be found in Lesson 6.

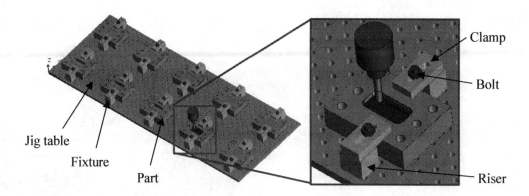

Figure 1.7 The material removal simulation of a multipart machining example

1.4   User Interface

The overall design of CAMWorks user interface, as shown in Figure 1.9, that includes the layout and windows, buttons, menu selections, dialog boxes, etc., is similar to that of SOLIDWORKS. SOLIDWORKS users should find it is straightforward to maneuver in CAMWorks. As shown in Figure 1.9, the user interface window of CAMWorks consists of pull-down menus, command buttons, graphics area, and feature manager window.

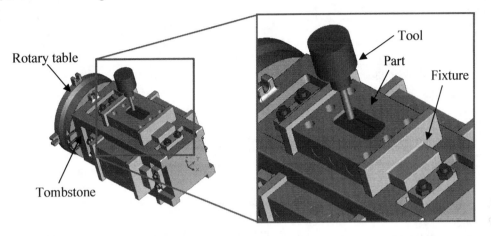

Figure 1.8   Material removal simulation of the multiplane machining example

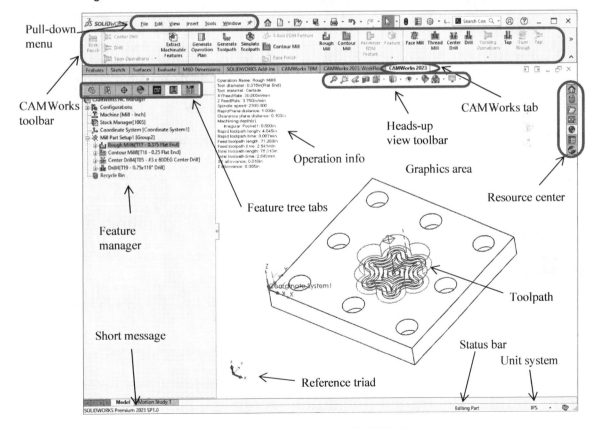

Figure 1.9   User interface of CAMWorks

Table 1.1 The major command buttons in CAMWorks

Button Symbol	Name	Function
Define Machine	Define Machine	Allows you to define the machine tool that the part will be machined on, such as 3-axis mill.
Coordinate System	Define Coordinate System	Allows you to define a coordinate system and assign it as the Fixture Coordinate System for the active machine.
Stock Manager	Stock Manger	Allows you to define the raw stock from a bounding box, an extruded sketch or an STL file.
Setup	Part Setup	Allows you to create Part Setups that defines (1) tool orientation (or feed direction), (2) G-code program zero, and (3) the X direction of tool motion.
Extract Machinable Features	Extract Machinable Features	Initiates automatic feature recognition (AFR) to automatically extract solid features that correspond to the machinable features defined in the technology database (TechDB™). The types of machinable features recognized for mill and turn are different. CAMWorks determines the types of features to recognize based on the NC machine selected. The machinable features extracted are listed in the feature manager window under the CAMWorks feature tree tab ▣ .
Generate Operation Plan	Generate Operation Plan	Generates operation plans automatically for the selected machinable features. The operation plans and associated machining strategy and machining parameters are selected based on rules defined in TechDB™. An operation plan contains information on how the machinable features are to be machined. The operations generated are listed in the feature manager window under the CAMWorks operation tree tab ◩ .
Generate Toolpath	Generate Toolpath	Creates toolpath for the selected operation plans and displays the toolpath on the part. A toolpath is a cutting entity (line, circle, arc, etc.) created by a cutting cycle that defines tool motion.
Simulate Toolpath	Simulate Toolpath	Provides a visual verification of the machining process for the current part by simulating the tool motion and the material removal process.
Step Thru Toolpath	Step Through Toolpath	Allows you to view toolpath movements either one movement at a time, a specified number of movements or all movements.
Save CL File	Save CL File	Allows you to save the current operation and associated parameters in the technology database as CL (cutter location) data for future use.
Post Process	Post Process	Translates toolpath and operation information into G-code for a specific machine tool controller.

An example file, *Lesson 1 with Toolpath.SLDPRT*, is prepared for you to browse numerous capabilities and become familiar with selections, buttons, commands and options of CAMWorks user interface. This file (and all example files of the book) is available for download at the publisher's website (www.sdcpublications.com). You may review Section 1.5 for steps to bring the example into CAMWorks.

The graphics area displays the solid or machining simulation model with which you are working. The pull-down menus provide basic solid modeling functions in SOLIDWORKS and machining functions in CAMWorks. The command buttons of the CAMWorks tab above the graphics area offer all the functions required to create and modify virtual machining operations in a generic order. Major buttons include extract machinable features, generate operation plan, generate toolpath, simulate toolpath, step through toolpath, save CL file, and post process. When you move the mouse pointer over these buttons, a short message describing the menu command will appear. Some of the frequently used buttons in CAMWorks and their functions are also summarized in Table 1.1 for your reference.

There are four feature tree tabs on top of the feature manager window that are highly relevant in learning CAMWorks. The left most tab, FeatureManager design tree ![icon] (see Figure 1.10), sets the display to the SOLIDWORKS design tree (also called model tree or solid feature tree), which lists solid features, parts, assembly mates, and subassemblies created in SOLIDWORKS in the feature manager window.

The third tab from the right, CAMWorks feature tree ![icon] (see Figure 1.11), shifts the display to the CAMWorks feature tree, which lists machinable features extracted or identified from the solid model. The tree initially shows only *Configurations*, *Machine* (for example, *Mill-in* in Figure 1.11), *Stock Manager*, *Coordinate System*, and *Recycle Bin*. The *Machine* node indicates the current machine as mill, turn, mill-turn, or wire EDM. You will have to select a correct machine before you begin working on a part. If you click any machinable feature, an outline view of the machinable feature appears in the part in the graphics area. For example, the sketch of the pocket appears when clicking *Irregular Pocket1* in the feature tree, as shown in Figure 1.11. Note that a symbol ![icon] (called tool axis symbol) appears indicating the tool axis direction (or feed direction) of all the machinable features under the current mill part setup.

The second tab from the right, CAMWorks operation tree ![icon] (shown in Figure 1.12), sets the display to the CAMWorks operation tree. After you select the *Generate Operation Plan* command, the operation tree lists NC operations for the corresponding machinable features.

Figure 1.10 Solid features listed in the feature manager window under the FeatureManager design tree

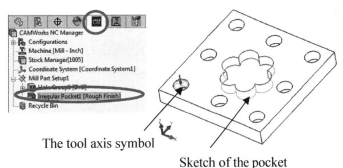

The tool axis symbol

Sketch of the pocket

Figure 1.11 Selecting a machinable feature under the CAMWorks feature tree tab

Similar to SOLIDWORKS, right clicking an operation in the operation tree will bring up command options that you can choose to modify or adjust the machining operation, such as feedrate, spindle speed, and so on. Clicking any operations after selecting the *Generate Toolpath* command will bring out the corresponding toolpaths in the part in the graphics area, like that of Figure 1.4.

The right most tab, CAMWorks tools tree ▣ (shown in Figure 1.13), sets the display to the CAMWorks tools tree. CAMWorks tools tree lists tools available in the tool crib you selected for the machine.

Figure 1.12 CAMWorks
operation tree tab

Figure 1.13 CAMWorks tools tree tab

1.5   Opening Lesson 1 Model and Entering CAMWorks

A machining simulation model for the simple block example shown in Figure 1.2 has been created for you. You may download all example files from the publisher's website, unzip them, and locate the model under Lesson 1 folder. Copy or move Lesson 1 folder to your hard drive.

Start SOLIDWORKS, and open model file *Lesson 1 with toolpath.SLDPRT*. You should see a solid model file like that of Figure 1.2 appearing in the graphics area.

Entering CAMWorks from SOLIDWORKS is straightforward. You may click the CAMWorks feature tree tab ▣ or operation tree tab ▣ to browse respective machining entities. You may right click any node listed in the feature or operation tree to modify or adjust the machining model. You may also choose options under the pull-down menu *Tools > CAMWorks* to launch the same commands of those listed in Table 1.1 (and more) that support you to extract machinable features, generate operation plan, and so on.

If you do not see the CAMWorks feature tree or operation tree tab, you may have not activated the CAMWorks add-in module. To activate the CAMWorks module, choose from the pull-down menu

Tools > Add-Ins

In the *Add-Ins* dialog box shown in Figure 1.14, click *CAMWorks 2023* in both boxes (*Active Add-ins* and *Start Up*), and then click *OK*. You should see that CAMWorks 2023 tab appears above the graphics area like that of Figure 1.9 and CAMWorks tree tabs added to the top of the feature manager window.

If you still do not see any of the CAMWorks tree tabs on top of the feature manager window or any of the CAMWorks buttons above the graphics area (like those of Table 1.1), you may have not set up your CAMWorks license option properly. To check the CAMWorks license setup, click the *Help* button ⊚ above the graphics area and select

CAMWorks 2023 > License Info

In the *CAMWorks License Info* dialog box (Figure 1.15), click all clickable boxes (or select modules as needed, for example, *3X Mill L1* for 3-axis mill level 1) to activate the module(s). Then click *OK*.

You may need to restart SOLIDWORKS to activate newly added CAMWorks modules. Certainly, before going over this tutorial lesson, you are encouraged to check with your system administrator to make sure SOLIDWORKS and CAMWorks have been properly installed on your computer.

Please note that if you have SOLIDWORKS CAM added in to your SOLIDWORKS, you will have to remove it by deselecting the module in the *Add-Ins* dialog box before adding in CAMWorks.

Another point worth noting is that the auto save option might have been turned on in CAMWorks by default. Often, it is annoying to get interrupted by this auto save every few minutes asking if you want to save the model. You may turn this auto save option off by choosing from the pull-down menu

Tools > CAMWorks > Options

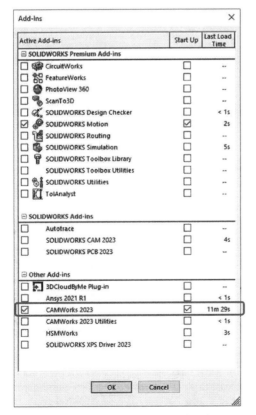

Figure 1.14 The *Add-Ins* dialog box

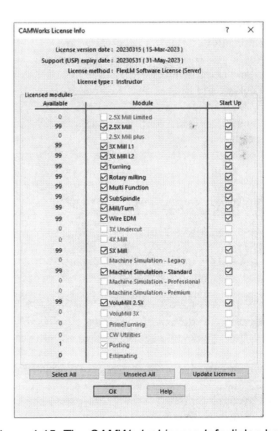

Figure 1.15 The *CAMWorks License Info* dialog box

In the *Options* dialog box (Figure 1.16), select *Disable Auto Saving* under the *General* tab to turn it off. Then click *OK*.

To browse an existing CAMWorks model, you may click any machinable features listed under the CAMWorks feature tree tab [CW] to display the feature in the graphics area. For example, the pocket profile sketch like that of Figure 1.11 appears in the design model after clicking *Irregular Pocket1*.

You may also click an operation under CAMWorks operation tree tab [icon] to show the toolpath of the selected operation; for example, *Rough Mill* or *Contour Mill* to display toolpaths like those of Figure 1.4. You may step through toolpath of an operation by right clicking it and choosing *Step Through Toolpath* (see Figure 1.5). You may right click an operation and choose *Simulate Toolpath* to simulate a material removal process of the operation like that of Figure 1.6.

1.6 Extracting Machinable Features

Machining operations and toolpaths can be generated only on machinable features. A unique and appealing technical feature in CAMWorks is the automatic feature recognition (AFR) technology, which analyzes the solid features in the part and extracts mill features such as holes, slots, pockets and bosses; turn features such as outside and inside diameter profiles, faces, grooves and cutoff, and wire EDM features such as die openings.

The AFR technology helps in reducing the time spent by the designer to feed in data and select options related to creating machining simulation.

Figure 1.16 The *Options* dialog box

A set of machinable features for milling and turning operations that can be extracted by AFR are summarized in Appendix A. The associated machining strategies of individual machinable features can be found in Appendix B.

The *Extract Machinable Features* command [icon] initiates AFR capability. Depending on the complexity of the part, AFR can save considerable time in extracting 2.5 axis features, such as holes, pockets, slots, bosses, etc., either prismatic (with vertical walls) or tapered.

AFR cannot recognize every single feature on a complex part and does not recognize features beyond 2.5 axis. To machine these areas, you need to define machinable features manually using the interactive feature recognition (IFR) wizard. For example, you may define a *Multi Surface* feature manually by selecting faces to be cut and faces to avoid in the design model. More on this topic will be discussed in this book, for example, in Lesson 2: Simple Plate, and Lesson 4: Freeform Surface.

1.7   Technology Database

CAMWorks technology database, TechDB™, is a self-populating database which contains all the information about the machine, cutting tools and the parameters used by the operator, and rules of repetitive NC operations (called strategy in CAMWorks) for the respective machinable features. This database within CAMWorks can be customized easily to meet the user's and the shop floor's requirements. The database can be utilized for all the manufacturing processes viz. milling, turning, mill-turn and wire EDM. This database supports best practices at a centralized location in the tool room; thus, it eliminates the non-uniformity in practicing virtual and physical machining operations.

Using a set of knowledge-based rules, CAMWorks analyzes the machinable features to determine the machining process plan for a design model. In this approach, features are classified according to the number of possible tool approaching directions that can be used to machine them. The knowledge-based rules are applied to assure that you get desired cutting operations. The rules that determine machining operations for a respective machinable feature can be found in Appendix B, for both milling and turning operations.

The technology database is shipped with data that is considered generally applicable to most machining environments. Data and information stored in the database can be added, modified, or deleted to meet the user's specific needs in practice. More about accessing and modifying the database is discussed in Lesson 10: Die Machining Application.

1.8   CAMWorks Machine Simulation

CAMWorks Machine Simulation offers a realistic machine setting in a virtual environment, in which machining simulation may be conducted on a virtual replicate of the physical NC machine. Computer models of the machine, tilt rotary table, fixtures, tool, tool holder, stock and part are assembled to realistically represent a physical NC machine setting. In addition to simulating machining operations, the Machine Simulation capability provides tool collision detection in a more realistic setting.

Although CAMWorks Machine Simulation capability is offered as a separate licensed module, legacy machines, including a mill and a mill-turn, come with CAMWorks. For example, Figure 1.17 shows machining simulation using a legacy mill, *Mill_Tutorial*, in which the tool, tool holder, tilt table, rotary table, and a stock together with a machine coordinate system XYZ are displayed (see Figure 1.17). To the right, the machining operations and the corresponding G-code is listed in the upper and lower areas of *Move List*, respectively. On top, a default virtual machine (or simulator), *Mill_Tutorial*, is selected. Below are buttons that control the machining simulation run. More about the Machine Simulation capability and commands can be found in Lesson 7: Multiaxis Milling and Machine Simulation.

More importantly, this capability allows you to install a virtual CNC machine that is a virtual replicate of a physical machine available at your machine shop on your computer to carry out machining simulation. For example, a HAAS mill can be added to CAMWorks as a virtual CNC machine for support of machine simulation, as shown in Figure 1.18. In this case, you will have to acquire computer models and files of the physical CNC machine, and place these files at subfolders of C:\CAMWorksData. We are fortunate that we were able to acquire the models and files of a HAAS mill, place them at C:\CAMWorksData, and carry out machining simulations with a virtual replicate of the HAAS mill. We will discuss this capability in Lesson 10: Die Machining Application. Unfortunately, we are not allowed to share the models and files with the readers.

NC machine

Tool holder

Tool

Tilt table

Stock

Rotary table

NC operations

G-code

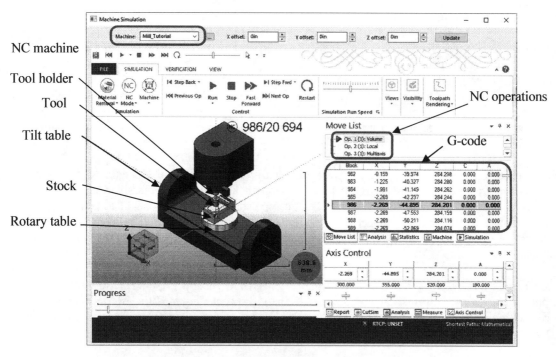

Figure 1.17 The *Machine Simulation* window of *Mill_Tutorial*

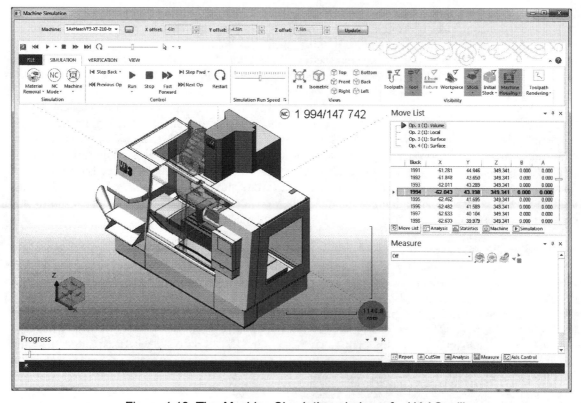

Figure 1.18 The *Machine Simulation* window of a HAAS mill

1.9   Tutorial Examples

In addition to the example of this lesson, nine machining examples are included in this book: seven milling and two turning. All nine lessons illustrate step-by-step details of creating machining operations and simulating toolpath capabilities in CAMWorks.

We start in Lesson 2 with a simple plate example, which provides you with a brief introduction to CAMWorks 2023 and offers a quick run-through for creating a contour (also called profile) mill operation using a 3-axis mill.

Lessons 3 through 7 focus on milling operations. We include examples of machining 2.5 axis features using 3-axis mill in Lesson 3, machining a freeform surface of a solid feature in Lesson 4, machining a set of identical parts in an assembly in Lesson 5, machining features on multiple part planes using a 3-axis mill with a rotary table in Lesson 6, and machining a cylindrical surface using multiaxis milling operation in Lesson 7. In Lesson 7 we also discuss Machine Simulation, in which we bring the machining simulation into a legacy mill that comes with CAMWorks software.

Lessons 8 and 9 focus on turning operations. In Lesson 8, we use a simple stepped bar example to learn basic capabilities in simulating turning operations and understanding G-code generated by CAMWorks. In Lesson 9, we machine a similar example with more turn features to gain a broader understanding of the turning capabilities offered by CAMWorks. In Lesson 9, we also discuss mill-turn operations.

In Lesson 10, we present an industrial application that involves die machining for sheet metal forming. In this application, tooling manufacturing for sheet metal forming, including punch and die, was carried out mainly by using a HAAS mill. CAMWorks was employed to conduct virtual machining and toolpath generation for the die and punch. The goal of the lesson is to offer readers a flavor of the role that CAMWorks would play in a practical tooling manufacturing application.

One thing we emphasize in this book is the verification of the G-code converted from virtual machining simulation. Learning the menu selections and button clicking of CAMWorks for generating machining operations is important. On the other hand, the virtual machining simulation must lead to something useful at the shop floor. That is, the G-code converted from machining operations must be accurate and compatible with the NC machines at the shop floor. The G-code must be ready and able to produce parts as desired without major hurdles. Please note that no software is 100% error-proof and bug-free. Therefore, it is extremely important that we carefully review and verify the G-code before loading it to the NC machine for material cutting. At the end of most lessons, we review and verify the G-code converted. Readers are strongly encouraged to do the same while applying the skills learned from this book to their own machining projects. Examples and topics to be discussed in individual lessons are summarized in Table 1.2.

Table 1.2 Examples employed and topics to be discussed in this book

Lesson	Example	Machining Model	Problem Type	Topics to Discuss
2	Simple Plate		3-axis contour (or profile) milling	1. A brief introduction to CAMWorks 2023 2. A complete process of using CAMWorks to create a milling operation from the beginning all the way to the post process that generates G-code 3. Extract machinable feature using interactive feature recognition (IFR) 4. Review and verify G-code generated
3	2.5 Axis Features		Pocket milling and hole drilling	1. Extract 2.5 axis machinable features using automatic feature recognition (AFR) 2. Identify machinable feature for face milling operation interactively 3. Define part setup origin for G-code generation 4. Adjust machining parameters to regenerate toolpath 5. Review and verify G-code generated
4	Freeform Surface		Machine a freeform surface using a 3-axis mill	1. Create a multi surface machinable feature and select avoid surface feature to restrain area of toolpath generation 2. Create *Area Clearance* (rough cut) and *Pattern Project* (finish cut) operations 3. Convert an area clearance operation to local milling operation 4. Take a closer look at the options offered for creating desirable *Pattern Project* operation 5. Create section views for a closer look at the freeform surface in material removal simulation
5	Multipart Machining		Machine a set of identical parts in an assembly	1. Select fixture coordinate system for cutting multiple parts 2. Create instances of part for machining operations 3. Define stocks for individual instances 4. Select components in the assembly for the tools to avoid 5. Simulate material removal for cutting multiple parts 6. Review and verify G-code generated

Table 1.2 Examples employed and topics to be discussed in this book (cont'd)

Lesson	Example	Machining Model	Problem Type	Topics to Discuss
6	Multiplane Machining		Machine features on multiple planes using a 3-axis mill with a rotary table	1. Cut parts with machinable features on multiple planes 2. Set the rotation axis of the rotary table as the 4th axis 3. Select components to avoid and to be included in material removal simulation 4. Rotate tool vs. rotate stock in material removal simulation 5. Review and verify G-code generated
7	Multiaxis Surface Machining		Multiaxis milling operations and Machine Simulation	1. Machine a cylindrical surface extruded by a Bézier curve using 5-axis mill 2. Create volume milling, local milling, and multiaxis surface milling operations 3. Identify tool gouging and regenerate toolpath to avoid gouging 4. Use Machine Simulation to simulate multiaxis machining operations in a setup with a tilt rotary table
8	Turning a Stepped Bar		Basic turning operations using 2-axis lathe	1. Follow a complete process in using CAMWorks to create a turning simulation from the beginning all the way to the post process that generates G-code 2. Interactively extract machinable features 3. Review and verify G-code generated
9	Turning a Stub Shaft		Advanced turning operations	1. Extract machinable features for turning, including face, groove, thread, and holes at both ends using AFR 2. Interactively create turn thread feature, generate toolpath and review and verify G-code generated 3. Choose mill-turn to machine the side cut features and a cross hole in the stub shaft
10	Die Machining Application		Practical application	1. Introduce an industrial application that involves die machining for sheet metal forming 2. Add a HAAS mill computer model to CAMWorks Machine Simulation 3. Customize the technology database 4. Use the added HAAS mill to support machine simulation 5. Add a HAAS post processor to CAMWorks for G-code generation 6. Collision detection in Machine Simulation

[Notes]

Lesson 2:   A Quick Run-Through

2.1   Overview of the Lesson

The purpose of this lesson is to provide readers with a quick start in using CAMWorks for carrying out virtual machining simulations. You will learn a complete process in using CAMWorks to create a machining simulation from the beginning all the way to the post process that generates G-code for carrying out physical CNC machining. We will use a 3-axis mill to machine a simple plate by carrying out a contour milling operation (often called profile milling in other CAM software) that cuts the boundary profile of the part, as shown in Figure 2.1. You may want to open the model file, *Simple Plate with Toolpath.SLDPRT*, to preview the machining operation generated for this lesson. This lesson is intentionally made simple for new CAMWorks users. We stay with default options and machining parameters for most of the selections.

Figure 2.1 The material removal simulation of the simple plate example

We follow the general steps shown in Figure 1.1 of Lesson 1 to create virtual machining simulation and generate toolpath using CAMWorks. In addition, we will go over a post process to convert G-code from this contour milling operation. We review and verify the accuracy of the G-code generated using a generic post processor of 3-axis mill that comes with CAMWorks at the end of this lesson.

After completing this lesson, you should be able to carry out virtual machining simulation and toolpath generation for similar applications that involve contour milling operations following the same process. This lesson should also prepare you for the remaining lessons of the book.

2.2   The Simple Plate Example

The L-shape plate has a bounding box of size 4in.×4in.×0.5in., as shown in Figure 2.2, with a fillet of 0.375in. in radius. The unit system chosen is IPS (inch, pound, second). There is one solid feature created in the design model, *Boss-Extrude1*, listed in the FeatureManager design tree 🌐 (or simply model tree or feature tree) shown in Figure 2.3. In addition, a coordinate system (*Coordinate System1*) is defined at the front left corner of the top face of the part. When you open the solid model *Simple Plate.SLDPRT*, you should see the solid feature and the coordinate system listed in the feature tree like that of Figure 2.3.

A stock of retangular block with a size 4.5in.×4.5in.×0.5in., made of low carbon alloy steel (1005), as shown in Figure 2.4, is chosen for the machining operation. Note that a part setup origin is defined at the front left vertex of the top face of the stock (instead of the coordinate system, *Coordinate System1*), which locates the G-code program zero.

We will create one contour milling operation, which cuts along the boundary profile of the part using a 1-in. flat-end mill. We will create the machinable feature interactively and follow the recommendations of the technology database (TechDB™) for choosing machining operation and parameters, such as feedrate and spindle speed. We will make a few minor changes to obtain a toolpath of the contour mill operation like that of Figure 2.5, in which the cutter moves along the boundary profile of the part in two passes. This is due to the fact that the machining depth (or depth of cut), 0.3in., chosen is less than the thickness of the stock (0.5in.).

2.3   Using CAMWorks

Open SOLIDWORKS Part

Open the part file (filename: *Simple Plate.SLDPRT*) downloaded from the publisher's website. This solid model, as shown in Figure 2.2, consists of one solid feature and a coordinate system among other entities. As soon as you open the model, you may want to check the unit system chosen.

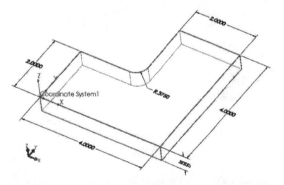

Figure 2.2 Dimensions of the simple plate model

Figure 2.3 Entities listed in the feature tree

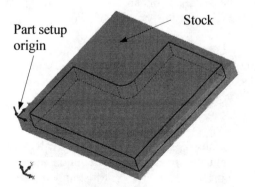

Figure 2.4 Stock with a part setup origin at its top left vertex

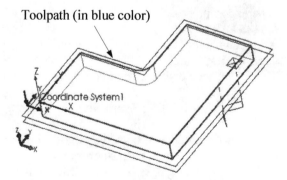

Figure 2.5 The toolpath of the contour milling operation

You may select from the pull-down menu *Tools > Options*. In the *Document Properties* dialog box (Figure 2.6), select the *Documents Properties* tab and select *Units*. Select *IPS (inch, pound, second)*. We will stay with this unit system for this lesson. Also, it is a good practice to increase the decimals from the default 2 to 4 digits since some machining parameters defined are down to a thousandth of an inch in CAMWorks. To increase the decimals to 4 digits, you may click the dropdown button to the right of the cell in the *Length* row under the *Decimals* column (circled in Figure 2.6). Pull down the selection and choose *.1234* for 4 digits. Click *OK* to accept the change.

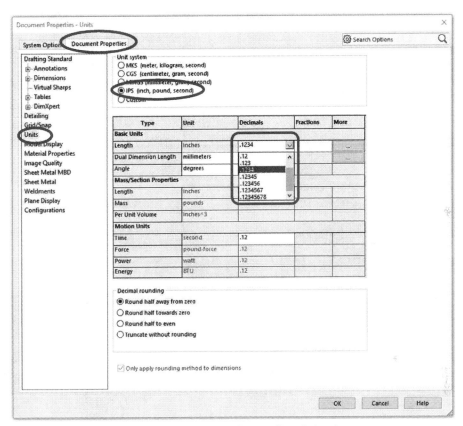

Figure 2.6 The *Document Properties* dialog box

Enter CAMWorks

At the top of the feature tree, you should see CAMWorks tabs, including CAMWorks feature tree ![CW] and CAMWorks operation tree ![C], as shown in Figure 2.7. You may click these two tabs to review machinable features and machining operations, respectively. You should also see CAMWorks command buttons like those of Table 1.1 above the graphics area (by clicking the *CAMWorks 2023* tab above the graphics area).

We will manually create a machinable feature (instead of clicking the *Extract Machinable Features* button ![Extract Machinable Features]), and use the four buttons next to it, *Generate Operation Plan* ![Generate Operation Plan], *Generate Toolpath* ![Generate Toolpath], *Simulate Toolpath* ![Simulate Toolpath], and *Step Through Toolpath* ![Step Thru Toolpath], to create and review the toolpath. Before going through these steps, we will first select an NC machine and define a stock to be employed for machining the part.

Select NC Machine

Click the CAMWorks feature tree tab . A default machine *Mill-inch*, which is a 3-axis mill of inch system, is listed in the feature manager window (see Figure 2.8). This is the NC machine we want to use. We right click *Mill-inch* and choose *Edit Definition*.

CAMWorks feature   CAMWorks operation
tree tab   tree tab

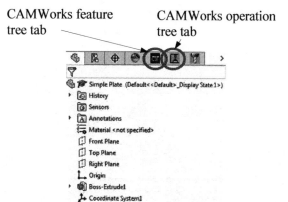

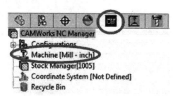

Figure 2.8 The NC machine
Mill-inch chosen by default

Figure 2.7 The CAMWorks feature tree and
CAMWorks operation tree tabs

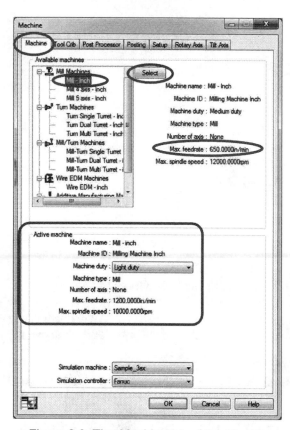

Figure 2.9 The *Machine* tab of the *Machine*
dialog box

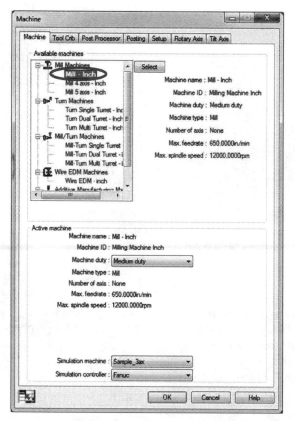

Figure 2.10 The *Mill-inch* machine selected and
highlighted in bold face

In the *Machine* dialog box (see Figure 2.9), available machines are listed in the box below *Available machines* under *Machine* tab. *Mill-inch* is listed under *Mill Machines*. A few basic machine parameters of the mill *Mill-inch* (for example *Max. feedrate*) are also listed. This is because a default *Mill-inch* of light duty is active (see information under *Active machine*).

Choose *Mill-inch* in the box under *Available machines*, and click *Select*. *Mill-inch* is now highlighted in bold face, as shown in Figure 2.10. Note that a medium duty machine is chosen.

Next, we will choose *Tool Crib*, *Post Processor*, and *Setup* tabs to review or modify the machine definition.

Choose the *Tool Crib* tab and select *Tool Crib 2* under *Available tool cribs* (see Figure 2.12), and then click *Select*. Tools available in *Crib 2* are now listed. We will use the tools available in *Crib 2* for this lesson.

Choose the *Post Processor* tab; a post processor called *M3AXIS-TUTORIAL* is selected (see Figure 2.11). This is a generic post processor of 3-axis mill that comes with CAMWorks.

There are other post processors that come with CAMWorks, which are located in *C:\ProgramData\SOLIDWORKS\CAMWorks 2023\Posts*. Note that in practice you will have to identify a suitable post processor that produces G-code compatible with the CNC machines at the shop floor. We will stay with *M3AXIS-TUTORIAL* for this lesson.

Figure 2.11 The *Tool Crib* tab of the *Machine* dialog box

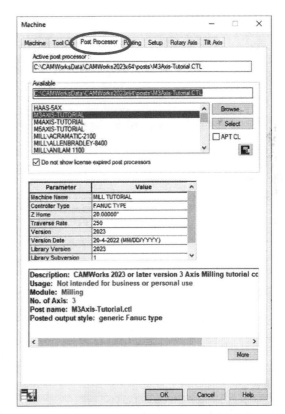

Figure 2.12 The *Post Processor* tab of the *Machine* dialog box

Choose the *Setup* tab, and click the *Define* button; see Figure 2.13(a). In the *Fixture Coordinate System* dialog box, choose *SOLIDWORKS Coordinate System*, circled in Figure 2.13(b), and click *Coordinate System1*. The selected coordinate system is now listed under *Selected Coordinate System*; see Figure 2.13(b).

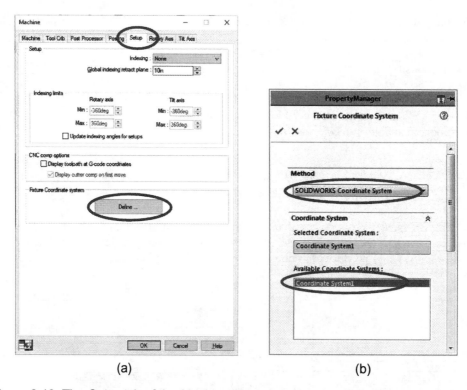

(a)   (b)

Figure 2.13 The *Setup* tab of the *Machine* dialog box, (a) choosing *Setup* tab, and clicking *Define*, and (b) choosing *Coordinate System1* for *Fixture Coordinate System*

Note that in CAMWorks, the fixture coordinate system defines the "home point" or main zero position on the machine. For some machining conditions, defining the fixture coordinate system is optional. However, to be safe, it is recommended that you always define an FCS.

Click *OK* to accept the selections and close the dialog box.

Create Stock

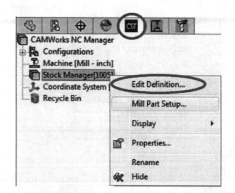

From CAMWorks feature tree , right click *Stock Manager* and choose *Edit Definition* (see Figure 2.14). The *Stock Manager* dialog box appears (Figure 2.15), in which a default stock size appears at the bottom of the dialog box, which is the size of the bounding box of the part. The default stock material is *1005* (if not, select *1005* for stock material). We will increase the length and width of the stock by 0.25in. on both sides by entering 0.25 for $X+$, $X-$, $Y+$, and $Y-$, as shown in Figure 2.15. The stock size becomes X:4.5, Y:4.5, Z:0.5, as circled in the *Stock Manager* dialog box (Figure 2.15).

Figure 2.14 Right clicking *Stock Manager* and choosing *Mill Part Setup* from the feature tree

Accept the revised stock by clicking the checkmark ✔ at the top left corner. The rectangular stock should appear in the graphics area similar to that of Figure 2.4.

Mill Part Setup and Machinable Feature

Since a machinable feature of the contour milling is not extractable by CAMWorks automatically, we will create one manually. We will first create a mill part setup and then insert a 2.5 axis feature.

From the CAMWorks feature tree ▣, right click *Stock Manager* and choose *Mill Part Setup*. The *Mill Setup* dialog box appears (Figure 2.16) is symbol ↑ with arrow facing upward appears. This symbol indicates that the tool axis (or the feed direction) must be reversed (pointing downward).

Click the *Reverse Selected Entity* button ↗ under *Entity* (circled in Figure 2.16) to reverse the direction. Make sure that the arrow points in a downward direction like ↧. Click the checkmark ✔ to accept the definition. A *Mill Part Setup1* is now listed in the CAMWorks feature tree, as shown in Figure 2.18.

Now we define a machinable feature. From CAMWorks feature tree ▣, right click *Mill Part Setup1* just created and choose *2.5 Axis Feature*.

In the *2.5 Axis Feature* dialog box (Figure 2.19), choose *Boss* for *Type*, and select *Sketch1* under *Available Sketches*, then click the *Next* button ● (circled in Figure 2.19) to define end condition.

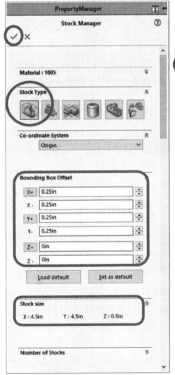

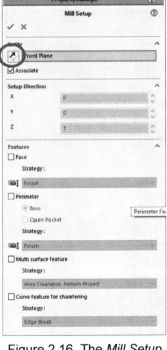

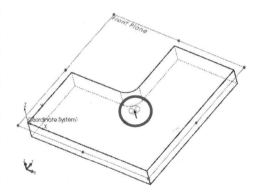

Figure 2.17 The *Front Plane* selected

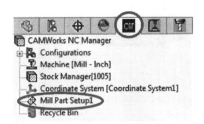

Figure 2.15 The *Stock Manager* dialog box

Figure 2.16 The *Mill Setup* dialog box

Figure 2.18 *Mill Part Setup1* listed in the CAMWorks feature tree

Note that *Sketch1* defines the boundary profile of the plate. Choose *Finish* (default) for *Strategy* (circled in Figure 2.20), and enter *0.5* for the depth dimension under *End condition – Direction 1*, then click the checkmark ✔ to accept the definition. An *Irregular Boss1* entity is now listed in the CAMWorks feature tree [CW] in magenta color (Figure 2.21), indicating that the machining feature is unfinished.

Generate Operation Plan and Toolpath

Right click *Irregular Boss1* and choose *Generate Operation Plan* (or click the *Generate Operation Plan* button ⬛ Generate Operation Plan above the graphics area). A new node, *Contour Mill1*, is listed in CAMWorks operation tree [icon] (see Figure 2.22) in magenta color. Note that you are now shifted automatically to CAMWorks operation tree [icon].

The NC operation, *Contour Mill1*, is assigned by CAMWorks by identifying a prescribed operation in the TechDB™, in which a 0.5in. diameter flat-end cutter is chosen.

Right click *Contour Mill1* and choose *Generate Toolpath* (or click the *Generate Toolpath* button [icon] Generate Toolpath above the graphics area).

A contour mill toolpath will be generated like that shown in Figure 2.23. Note that the flat-end cutter cuts along the part profile in two passes; the depth of cut smaller than the stock thickness was chosen by CAMWorks.

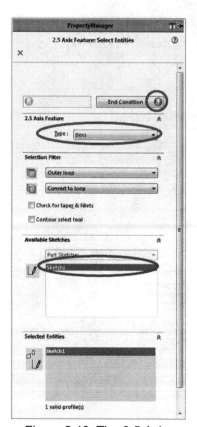

Figure 2.19 The *2.5 Axis
Feature* dialog box

Figure 2.20 Defining end
conditions

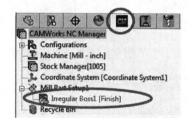

Figure 2.21 Machinable feature
added to the feature tree

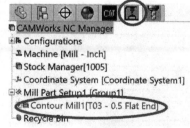

Figure 2.22 Machining operation
added to the operation tree

Define Part Setup Origin

The toolpath generated, as shown in Figure 2.23, assumes a part setup origin (see the symbol ⤷ circled in Figure 2.23) located at the front left corner of the design part, coinciding with the fixture coordinate system, *Coordinate System1*. You may see the part setup origin symbol more clearly by right clicking the coordinate system node, *Coordinate System1*, from the FeatureManager design tree 🌳 and choosing *Hide* 🚫 to hide it.

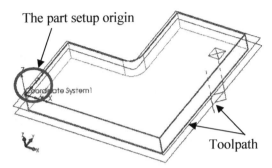

The part setup origin

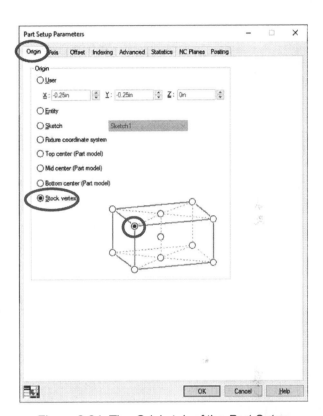

Figure 2.23 The toolpath generated

As mentioned before, the part setup origin defines the program zero for the G-code to be generated after the toolpath is finalized. A part setup origin defined at the corner of the design part may not be desirable. In practice, it is desirable to set the origin at a corner of the stock, for example, its top left corner, which is accessible and easier to set up on the mill.

We right click *Mill Part Setup* under the CAMWorks operation tree tab 🔧 and choose *Edit Definition*. In the *Part Setup Parameters* dialog box (Figure 2.24), select the *Origin* tab, choose *Stock vertex* and pick the vertex at the front corner of the top face of the sample stock (circled in Figure 2.24).

Figure 2.24 The *Origin* tab of the *Part Setup Parameters* dialog box

The part setup origin in the graphics area is now moved to the front left corner of the top face of the stock like that of Figure 2.4. Click *OK* to accept the change.

Click the *Yes* button in the CAMWorks warning box: *The origin or machining direction or advanced parameters has changed, toolpaths need to be recalculated. Regenerate toolpaths now?* The toolpath will be regenerated referring to where the setup origin is relocated.

Modify the Toolpath

Next we learn how to modify an existing operation.

In this case, we edit *Contour Mill1* by replacing the tool with a 0.75in. flat-end mill, and increasing the machining depth to 0.3in.

Under the CAMWorks operation tree tab ![icon], right click *Contour Mill1* and choose *Edit Definition*. In the *Operation Parameters* dialog box, choose *Tool* tab; the 0.5in. flat-end cutter is shown (Figure 2.25). Click the *Tool Crib* tab, select the 4th tool (Type: *FLAT END*, ID: *76*, Comment: *3/4 EM CRB 2FL 1-1/2LOC*, implying 3/4in. End Mill of Solid Carbide with 2 Flutes and 1.5in. Length of Cut), click the *Select* button to select the tool (see Figure 2.26).

Click *Yes* to the question in the warning box: *Do you want to replace the corresponding holder also?*

Choose the *Contour* tab of the *Operation Parameters* dialog box. In the *Depth* parameters group, click the percentage button ![icon] to deselect it for both *First cut amt.* and *Max cut amt.* Enter *0.3in* for both *First cut amt.* and *Max cut amt.*, as shown in Figure 2.27, since a larger tool has been chosen.

Click the *F/S* tab to review machining parameters (Figure 2.28); for example, the *XY feedrate* is chosen as *4.437in./min.* by the technology database of CAMWorks. Click *OK* to accept the changes.

Right click *Contour Mill1* node and choose *Generate Toolpath*. Profile milling toolpath will be generated like that shown in Figure 2.5 with two passes.

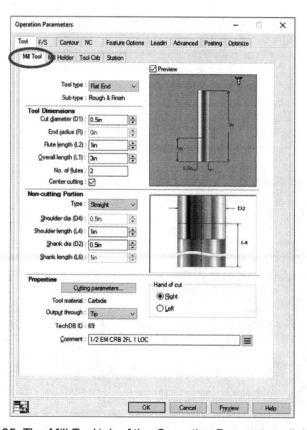

Figure 2.25 The *Mill Tool* tab of the *Operation Parameters* dialog box

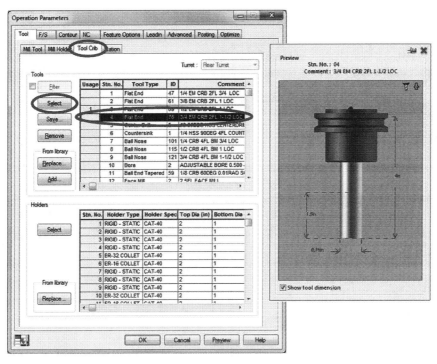

Figure 2.26 Selecting a 0.75in. flat-end cutter under the *Tool Crib*

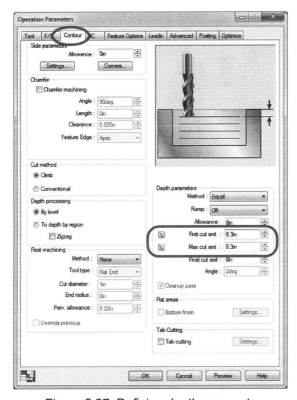

Figure 2.27 Defining depth parameters under the *Contour* tab

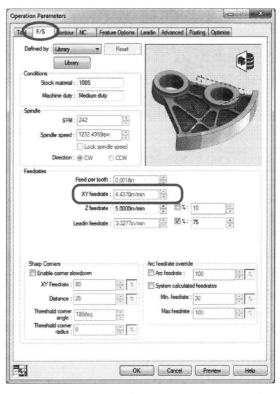

Figure 2.28 The *XY feedrate* shown under the *F/S* tab

Simulate Toolpath

Right click *Contour Mill1* node and choose *Simulate Toolpath* (or click the *Simulate Toolpath* button above the graphics area). The *Toolpath Simulation* tool box appears (Figure 2.29).

Click the *Run* button ▶ circled in Figure 2.29 to simulate the toolpath via a material removal simulation. The machining simulation of the contour milling operation will appear in the graphics area, similar to that of Figure 2.1.

Step Through Toolpath

You may step through the toolpath to learn more about the individual machining steps. Right click *Contour Mill1* node and choose *Step Thru Toolpath* (or click the *Step Thru Toolpath* button ⊟ Step Thru Toolpath above the graphics area). The *Step Through Toolpath* dialog box appears (Figure 2.30). Under *Information*, CAMWorks shows the tool movement from the current to the next steps (in X, Y, and Z coordinates), the feedrate and spindle speed, among others.

Click *Step* button ⏭ at the center of the tool box (circled in Figure 2.30) to step through the toolpath. You may want to turn on *Show toolpath points*, and *Tool Holder Shaded Display* (circled in Figure 2.30) to see the toolpath displayed on the part similar to that of Figure 2.31.

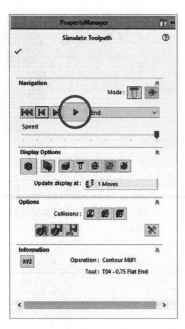

Figure 2.29 Playing the material removal
simulation by clicking the *Run* button

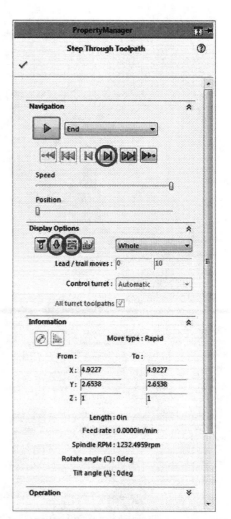

Figure 2.30 The *Step
through Toolpath* dialog box

2.4 The Post Process and G-Code

Now we have completed the contour milling operation. Next, we learn how to convert the toolpath to G-code and how to create a CL data file.

Right click *Contour Mill1* node and choose *Post Process* (or click the *Post Process* button ⟨icon⟩ above the graphics area). In the *Post Output File* dialog box (Figure 2.32), choose a proper file folder, enter a file name (default name is *Simple Plate.txt*), use the default type (*M3Axis-Tutorial*), and click *Save*. The *Post Process* dialog box appears (Figure 2.33).

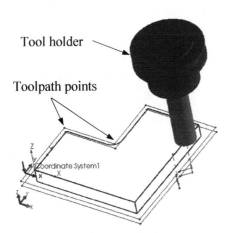

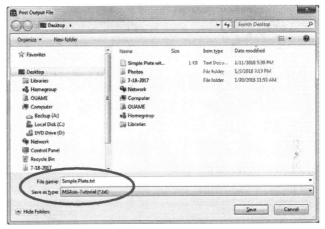

Figure 2.31 Stepping through toolpath

Figure 2.32 The *Post Output File* dialog box

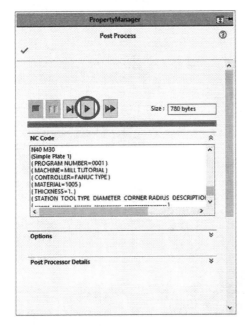

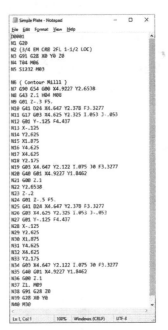

Figure 2.33 The *Post Process* dialog box

Figure 2.34 The contents of the *Simple Plate.txt* file

In the *Post Process* dialog box, click the *Play* button ▶ (circled in Figure 2.33) to create a G-code file (.txt file).

Open the G-code file (filename: *Simple Plate.txt*) from the folder using *Word* or *Word Pad* (see the file contents shown in Figure 2.34 with command explanations provided in Column B of Table 2.1).

Note that the NC word G41 in blocks N10 and N25 turns on tool radius compensation from the left for Tool#24 (D24 is included), which moves the tool to the left of the program path (that is, the boundary profile of the design model) by the amount of the radius of Tool#24. Since the cutter locations of the G-code specify the actual locations of the tool already (such as Points A, B, C, etc. shown in Figure 2.38), tool radius compensation is not necessary. In fact, Tool#24 was never loaded. G41 D24 in Blocks N10 and N25 are not needed at all. This is an error created by the post processor, *M3AXIS-TUTORIAL*, which can be corrected.

After correcting the error (and perhaps a few more, depending on the controller of your NC machine), the .txt file can be uploaded to a CNC mill, for example a HAAS mill, to machine the part. In general, the G-code converted is ready to be loaded to a mill. However, a few minor changes may be needed (or errors found in this example to be corrected), depending on the post processor employed in outputting the G-code output.

Click the *Save CL File* button 🖳 Save CL File above the graphics area. The *Save As* dialog box appears (Figure 2.35).

In the *Save As* dialog box (Figure 2.35), choose a proper file folder, enter a file name (default name is *Simple Plate.clt*), use the default type (*CL Files*), and click *Save*. Open the CL file to review its contents (Figure 2.36). The GOTO indicates how the cutter moves along the profile of the part. The X-, Y-, and Z-coordinates of the cutter location are identical to those of the G-code.

2.5   Reviewing Machining Time

From CAMWorks operation tree 🖳 , right click *Contour Mill1* and choose *Edit Definition*.

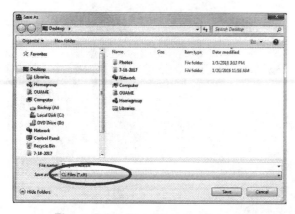

Figure 2.35 The *Save As* dialog box

```
UNITS/ INCHES
BOX/ 2.000000,2.000000,-0.250000,4.000000,4.000000,0.500000
CUTTER/ 0.750000,0.000000,0.375000,0.000000,0.000000,0.000000,4.000000,
TLAXIS/ 0.000000,0.000000,1.000000
OPFEATSTART/Contour Mill1-Irregular Boss1
ROTABL/ 0.000000,AAXIS,TABLE,CCLW
ROTABL/ -0.000000,BAXIS,TABLE,CCLW
ROTABL/ 0.000000,CAXIS,TABLE,CCLW
RAPID/
GOTO/ 4.922739,2.653805,1.000000
RAPID/
GOTO/ 4.922739,2.653805,0.100000
FEDRAT/ IPM,5.000000
GOTO/ 4.922739,2.653805,-0.300000
CUTCOM/ ON
CUTCOM/ LEFT
FEDRAT/ IPM,3.327739
GOTO/ 4.646967,2.378033,-0.300000
CIRCLE/ 4.700000,2.325000,-
0.300000,0.000000,0.000000,1.000000,0.075000,COUNTERCLOCKWISE
GOTO/ 4.625000,2.325000,-0.300000
FEDRAT/ IPM,4.436985
GOTO/ 4.625000,-0.125000,-0.300000
GOTO/ -0.125000,-0.125000,-0.300000
GOTO/ -0.125000,2.625000,-0.300000
GOTO/ 1.875000,2.625000,-0.300000
GOTO/ 1.875000,4.625000,-0.300000
GOTO/ 4.625000,4.625000,-0.300000
GOTO/ 4.625000,2.175000,-0.300000
FEDRAT/ IPM,3.327739
```

Figure 2.36 The partial contents of the
CL file (*Simple Plate.clt*)

Table 2.1 Contents of the .txt file (NC codes) with explanations

NC Commands (Column A)	Explanations (Column B)
O0001	Main program, program #0001
N1 G20	G20: Select inches
N2 G91 G28 X0 Y0 Z0	G91: incremental programming, G28: Return to reference point, move cutter to X0 Y0 Z0
N3 (Contour Mill1)	Comments in parentheses
N4 (3/4 EM CRB 2FL 1-1/2 LOC)	Comments in parentheses
N5 T04 M06	M06: Tool Change, Load tool #T04
N6 S1232 M03	M03: Spindle forward at speed 1232rpm
N7 G90 G54 G00 X4.9227 Y2.6538	G90: Absolute programming, G54: Select Work Coordinate System 1, G00: Move cutter to start point (X4.9227 Y2.6538) rapidly, non-cutting
N8 G43 Z.1 H04 M08	G43: Tool length compensation with length registered for the cutter #04, M08: Coolant on
N9 G01 Z-.3 F5.	G01: Cutting, plunge the cutter 0.3in. into the stock at feedrate 5. in/min
N10 **G41 D24** X4.647 Y2.378 F3.3277	G41: 2D cutter radius compensation from left *(this is an error, should be removed)*, D24: tool diameter registered at #24 *(this is an error since tool is T04, should be removed)*.
N11 G17 G03 X4.625 Y2.325 I.053 J-.053	G17: Cutting on the XY plane, G03: Cutting, circular motion moving the cutter to Point A (see Figure 2.38 for the toolpath points)
N12 G01 Y-.125 F4.437	G01: Cutting, move cutter to Point B at feedrate 4.437in/min
N13 X-.125	Point C
N14 Y2.625	Point D
N15 X1.875	Point E
N16 Y4.625	Point F
N17 X4.625	Point G
N18 Y2.175	Point A': X4.625 Y2.175 (not shown in Figure 2.38)
N19 G03 X4.647 Y2.122 I.075 J0 F3.3277	G03: Cutting, circular motion moving the cutter to X4.647 Y2.122
N20 G40 G01 X4.9227 Y1.8462	G40: Cancel radius compensation (no need), G01: Cutting, linear motion moving cutter to X4.9227 Y1.8462
N21 G00 Z.1	Retract to 0.1in. in the Z-direction, rapid
N22 Y2.6538	Move to X4.9227 Y2.6538
N23 Z-.2	Move to -0.2 in the Z-direction, rapid
N24 G01 Z-.5 F5.	G01: Cutting, plunge the cutter 0.5in. into the stock at feedrate 5.0 in/min for the second pass
N25 **G41 D24** X4.647 Y2.378 F3.3277	Same as block N10
N26 G03 X4.625 Y2.325 I.053 J-.053	Similar to Block N11 (Point A)
N27 G01 Y-.125 F4.437	Same as N12 (Point B)
N28 X-.125	Point C
N29 Y2.625	Point D
N30 X1.875	Point E
N31 Y4.625	Point F
N32 X4.625	Point G
N33 Y2.175	Point A'
N34 G03 X4.647 Y2.122 I.075 J0 F3.3277	Same as N19
N35 G40 G01 X4.9227 Y1.8462	Same as N20
N36 G00 Z.1	Retract to 0.1in. in the Z-direction, rapid
N37 Z1. M09	Retract to 1.0 in. in the Z-direction, rapid, M09: Coolant Off
N38 G91 G28 Z0	G91: Incremental programming, G28: Return to reference point, move cutter to Z0
N39 G28 X0 Y0	G28: Return to reference point, move cutter to X0Y0
N40 M30	Program end and rewind

In the *Operation Parameters* dialog box (Figure 2.37), choose the *Optimize* tab, and look for *Estimated machining time.* The total toolpath length of feed is 40.8in., and the total feed time is 9.31 minutes.

The feed time can be manually calculated by dividing the distance that the cutter travels (see Figure 2.38) by the feedrate.

Since the total distance *d* that the cutter travels along the part boundary profile is approximately 38in. (19in. per pass and there are two passes), as tabulated in Table 2.2, and the feedrate *f* is *4.437* in./min. (see Figure 2.28), the feed time can be calculated as

Feed Time = d/f = 38/4.437 = 8.56 minutes

The feed time calculated is not exactly the same as estimated by CAMWorks since the feed in the vertical direction (for examples, N9 and N24 in Table 2.1) and before cutter reaching Point A (N10, N11, N19, N25, N26, N34) were not included. However, this value is close to that of CAMWorks; i.e., 9.31 minutes shown in Figure 2.37.

We have completed this tutorial lesson. You may save your model for future reference.

Table 2.2 Cutter travel distance in one
pass with feedrate turned on

Point	Distance (in.)
AB	2.45
BC	4.75
CD	2.75
DE	2
EF	2
FG	2.75
GA	2.3
Total	19

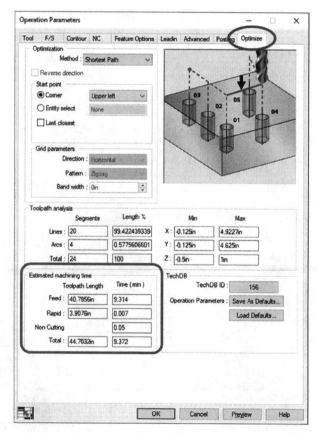

Figure 2.37 The *Optimize* tab of the
Operation Parameters dialog box

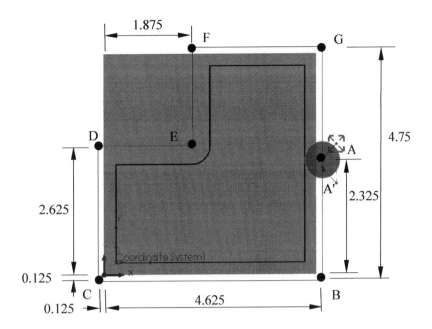

Figure 2.38 Dimensions of the cutter location points (labeled as A, B, C, and so on) and a shaded stock

2.6   Exercises

Problem 2.1.
In this lesson, we learned to create a 2.5 axis feature for the contour milling operation. In fact, there is another feature called *Part Perimeter Feature* that may also be suitable for this example in practice. You may create such a machinable feature by right clicking *Mill Part Setup1* and choosing *Part Perimeter Feature*. For this problem, you are asked to create such a part perimeter feature and carry out a machining simulation following the process learned in this lesson. Review the toolpath generated. Compare the toolpath with that of this lesson. Identify major differences, pros, and cons between these two. Which toolpath is more realistic?

Problem 2.2. Create a contour milling operation using CAMWorks to machine a block for the design model shown in Figure 2.39 by creating a 2.5 axis feature. The part file can be found at the publisher's website. It is located in Lesson 2/Exercises folder. The stock is 7in.×4.5in.×0.5in. of steel 1005, as shown in Figure 2.39. The feedrate is assumed 5 in./minute, and the depth of cut is 0.5in. In this exercise, please report the following:

- The tool you chose;
- Feed time obtained from CAMWorks;
- Feed time obtained from your own calculations;
- Screen captures for toolpath and material removal simulation in CAMWorks.

Grade should be given based on the following:

- Quality of the toolpath, i.e., the machined part must be identical to the design model (a clean cut);
- Feed time, a least possible feed time is desired.

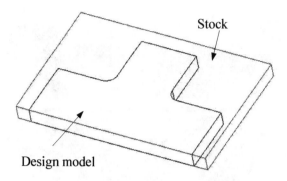

Figure 2.39 Design model and stock of Problem 2.2

Problem 2.3. Repeat Problem 2.2 for the design model shown in Figure 2.40. The stock is a rectangular block of 35in.×20in.×2in. of steel 1005. The feedrate is assumed 5 in./minute, and the step depth is 0.5 inch.

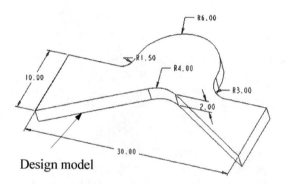

Figure 2.40 Design model of Problem 2.3

Lesson 3: Machining 2.5 Axis Features

3.1 Overview of the Lesson

In this lesson, we focus on learning milling operations for machining 2.5 axis features. The major characteristic of a 2.5 axis feature is that the top and bottom of the feature are flat and are normal to the tool axis of the machining operations under the mill part setup. Such features include prismatic solid features and solid features with tapered walls. Typical 2.5 axis features include boss, pocket, open pocket, corner slot, slot, hole, face feature, open profile, curve or engrave feature. Some of these features are illustrated in Figure A.1 of Appendix A.

Features of this type are often represented as a profile sketch in CAD software that gives the height of the feature at individual characteristic points. 2.5 axis features are often preferred for machining, as it is easy to generate G-code for them in an efficient, often close to optimal fashion. In general, 2.5 axis features can be machined using a 3-axis mill.

Most 2.5 axis features, such as pockets, holes, slots, and bosses, can be automatically extracted as machinable features by using the automatic feature recognition (AFR) capability of CAMWorks. Others, such as face milling features, or contour (or profile) milling features (as seen in Lesson 2), can be created manually (or interactively) using Interactive Feature Recognition (IFR) capability. In this lesson, we will use both methods, AFR and IFR, to extract and select respective machinable features from a solid model, generate operation plans and toolpaths, simulate and step through toolpaths, and post process the toolpaths for G-code.

The part solid model (or design model) of this lesson shown in Figure 3.1(a) consists of six holes and a pocket. They are all 2.5 axis features and are extracted as machinable features automatically. In addition, we will create a face milling operation that removes a thin layer of material on top of the part. A material removal simulation is shown in Figure 3.1(b).

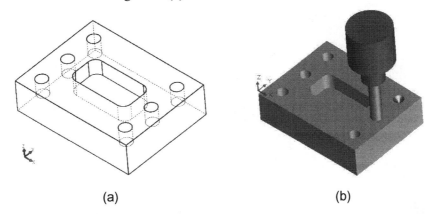

(a)   (b)

Figure 3.1 The 2.5 axis features example, (a) part solid model, and (b) material removal simulation

After completing this lesson, you should be able to generate machining simulations for similar parts that involve machining 2.5 axis features following the same steps. In this exercise, we will use default options for most of the selections.

3.2   The 2.5 Axis Features Example

The size of the bounding box of the part (filename: *2 point 5 axis features.SLDPRT*) is 8in.×6in.×2in., as shown in Figure 3.2(a). The base block is created as a *Boss-Extrude1* solid feature. The size of the center pocket, *Cut-Extrude1*, is 4in.×2in.×1in. with fillets of 0.5in. in radius at the four corners; see Figure 3.2 (b). There are six blind holes, three on each side, of diameter 0.75in. and depth 1.0in.; see Figure 3.2 (c). The first hole was created as a cut extrude feature (*Cut-Extrude2*) and then duplicated for additional instances by using a linear pattern feature, as shown in Figure 3.2 (d).

In addition, a coordinate system (*Coordinate System1*) is created at the front left corner of the bottom face of the part. This coordinate system will be chosen as the fixture coordinate system, which defines the "home point" or main zero position on the machine. However, please note that the G-code generated refers to the part setup origin, as discussed in Lesson 2, which does not have to be the same as the fixture coordinate system. The unit system chosen is IPS (inch, pound, second). When you open the solid model *2 point 5 axis features.SLDPRT*, you should see the four solid features and a coordinate system listed in the solid feature tree like that of Figure 3.3.

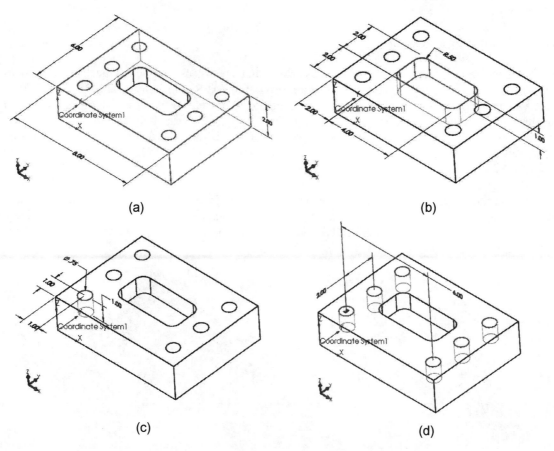

(a)	(b)
(c)	(d)

Figure 3.2 Dimensions of the solid features of the design model: (a) the base extrude, (b) the pocket, (3) the corner hole, and (d) the hole linear pattern

A stock of rectangular block with size 8in.×6in.×2.25in., made of low carbon alloy steel (1005), as shown in Figure 3.4, is chosen for the machining operations. Note that a part setup origin is defined at the front left vertex at the top face of the stock, which locates the G-code program zero. A layer of 0.25in. above the top face of the part will be removed by creating a face milling operation.

Figure 3.3 Feature tree of the example part

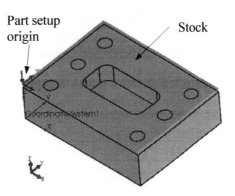

Figure 3.4 Stock with a part setup origin created at the front left vertex on top face

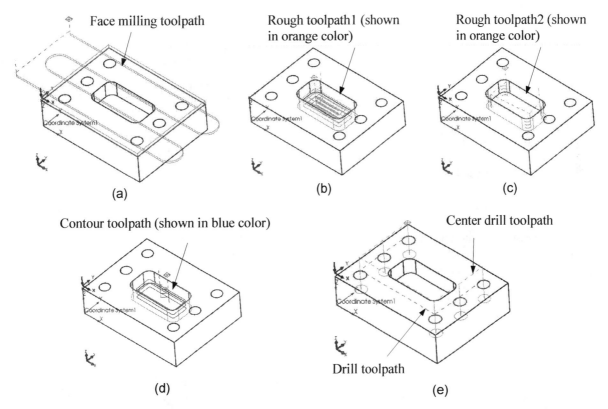

Figure 3.5 Toolpaths of the six NC operations: (a) face milling, (b) pocket milling: Rough Mill1, (c) pocket milling: Rough Mill2, (d) pocket milling: Contour Mill1, and (e) hole drilling: Center Drill1 and Drill1

There are six NC operations to be generated for this example. The first operation is face milling, which is a rough cut using a 2.5in. flat-end mill (T21 in the tool crib). The toolpath of the face milling is shown in Figure 3.5(a).

The second operation—see Figure 3.5(b)—is a rough mill operation that cuts the center pocket (*Cut-Extrude1* solid feature) using a 0.75in. flat-end mill (T04). The third—see Figure 3.5(c)—is another rough mill that cuts the fillets of the center pocket using a 0.5in. flat-end mill (T03). The fourth operation—see Figure 3.5(d)—is a contour mill for a finish cut along the inner boundary face of the pocket using the same 0.75in. flat-end mill (T04). The last two operations drill the six holes, which consist of center drill using a 3/4×90DEG center drill bit (T19) and hole drilling using a 0.75×135° drill bit (T20). The toolpath of the hole drilling operations (including the center drill) is shown in Figure 3.5(e).

The pocket and holes are recognized as machinable features using the automatic feature recognition (AFR), and the machinable feature for the face milling operation will be created interactively. Like Lesson 2, we will use mostly the default options and parameters chosen by TechDB™.

You may open the example file with toolpaths created (filename: *2 point 5 axis features with toolpath.SLDPRT*) to preview its toolpaths. When you open the file, you should see the six operations listed under the CAMWorks operation tree tab 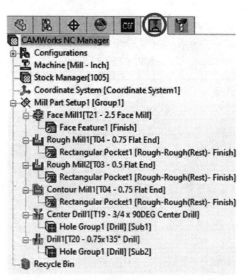, as shown in Figure 3.6. You may simulate individual operations by right clicking the operation and choosing *Simulate Toolpath*. You may simulate combined operations by selecting them (press the control key) and pressing the right mouse button to select *Simulate Toolpath*. You may also click *Mill Part Setup1* and choose the *Simulate Toolpath* button ⬢ above the graphics area to see material removal simulation for all six operations combined.

Figure 3.6 The six NC operations listed under the CAMWorks operation tree tab

3.3   Using CAMWorks

Open SOLIDWORKS Part

Open the part solid model (filename: *2 point 5 axis features.SLDPRT*) downloaded from the publisher's website. This solid model, as shown in Figure 3.2, consists of four solid features and a coordinate system. As soon as you open the model, you may want to check the unit system chosen and make sure the IPS system is selected. You may also increase the decimals from the default 2 to 4 digits following the steps similar to those discussed in Lesson 2.

Enter CAMWorks

Like what we learned in Lesson 2, at the top of the feature tree you should see CAMWorks tree tabs, including CAMWorks feature tree 🗂 and CAMWorks operation tree 🗎 . You should also see CAMWorks command buttons in the ribbon bar above the graphics area (by clicking the *CAMWorks 2023* tab above the graphics area; see Figure 1.9 of Lesson 1).

Select NC Machine

Click the CAMWorks feature tree tab 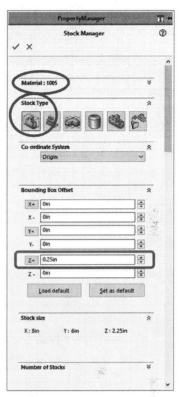 and right click *Mill-inch* to select *Edit Definition*. Similar to those of Lesson 2, in the *Machine* dialog box, we select *Mill-inch* under *Machine* tab, choose *Tool Crib2* under *Available tool cribs* of the *Tool Crib* tab, select *M3AXIS-TUTORIAL* under the *Post Processor* tab, and select *Coordinate System1* under *Fixture Coordinate System* of the *Setup* tab.

Create Stock

From CAMWorks feature tree ▉ , right click *Stock Manager* and choose *Edit Definition*. In the *Stock Manager* dialog box (Figure 3.7), we increase the height of the stock by 0.25in. on top (that is, enter *0.25* for *Z+*, circled in Figure 3.7). Choose *1005* for stock material. Accept the revised stock by clicking the checkmark ✔ at the top left corner of the dialog box. A rectangular stock should appear in the graphics area similar to that of Figure 3.4.

Mill Part Setup and Machinable Features

Click the *Extract Machinable Features* button ⬚ above the graphics area (or choose from the pull-down menu *Tools > CAMWorks > Extract Machinable Features*). A *Mill Part Setup1* is created with two machinable features extracted: *Rectangular Pocket1* and *Hole Group1* (with six holes), all listed in CAMWorks feature tree ▉ (Figure 3.8). All machinable features are shown in magenta color since there are no NC operations generated for these features yet.

Generate Operation Plan and Toolpath

Click the *Generate Operation Plan* button ⬚ above the graphics area. Five NC operations: 2 rough mills, 1 contour mill, 1 center drill, and 1 drill, are generated. They are listed in CAMWorks operation tree ▉ , as shown in Figure 3.9.

Figure 3.7 The *Stock Manager* dialog box

Figure 3.8 The machinable features extracted

Figure 3.9 The machining operation plans generated

Again, these features are shown in magenta color, indicating that these operations are not completely defined yet. Click the *Generate Toolpath* button 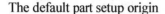 above the graphics area to create toolpath. The five operations are turned into a black color after toolpaths are generated.

Note that the default part setup origin coincides with the origin of the coordinate system (*Coordinate System1*), as circled in Figure 3.10, which is not desired for this example. As discussed in Lesson 2, the part setup origin defines the program zero location for the G-code. We relocate the origin to the front left corner at the top face of the stock. This corner point of the stock is easier to access at a CNC mill and is commonly chosen as the program zero in carrying out machining tasks. You may certainly select any adequate location as the G-code origin as long as the selected origin can be physically set up on the workbench or jig table of the mill at the shop floor.

The default part setup origin

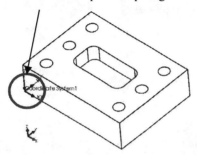

Figure 3.10 The default part setup origin coinciding with *Coordinate System1*

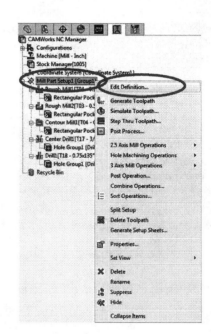

Figure 3.11

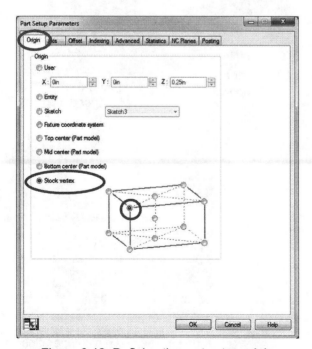

Figure 3.12 Defining the part setup origin

Part setup origin

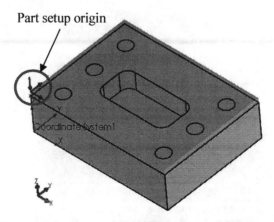

Figure 3.13 The part setup origin properly relocated

Relocate the origin by right clicking *Mill Part Setup1* under the CAMWorks operation tree 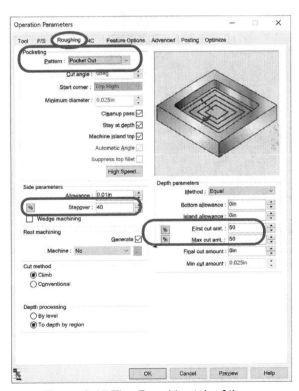 and select *Edit Definition* (see Figure 3.11).

In the *Part Setup Parameters* dialog box (Figure 3.12), choose *Stock vertex* (under the *Origin* tab), and pick the vertex at the front corner of the top face of the stock, as circled in Figure 3.12. Click *OK* to accept the change. A warning box appears. Click *Yes* to the question in the warning box: *The origin or machining direction or advanced parameters has changed, toolpaths need to be recalculated. Regenerate toolpaths now?*

The toolpaths will be regenerated, and the part setup origin is now moved to the top left corner of the stock in the graphics area, as shown in Figure 3.13.

You may review individual operations listed under CAMWorks operation tree 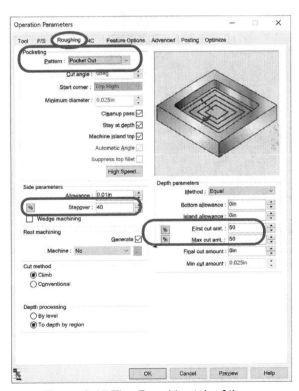 (see Figure 3.9), for example, *Rough Mill1*. Right click *Rough Mill1* and choose *Edit Definition*. In the *Operation Parameters* dialog box (Figure 3.14), *3/4 EM CRB 2FL 1-1/2 LOC* (that is, 0.75in. diameter flat-end mill, carbide, 2 flutes, and 1.5 in. length-of-cut) is chosen. Click the *Roughing* tab; the dialog box (Figure 3.15) shows *Pocket Out* as the pocketing pattern, stepover: 40% of tool diameter, and depth parameters: 50% of the tool diameter for both the *First cut amt.* and *Max cut amt.*

You may choose other tabs, such as *F/S* (feedrate and spindle speed), to review the machining parameters determined by the TechDB™. You may change these parameters and regenerate the toolpath as desired. We will simply click *Cancel* to close the dialog box.

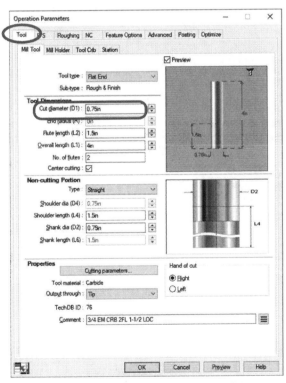

Figure 3.14 The *Mill Tool* tab of the *Operation Parameters* dialog box

Figure 3.15 The *Roughing* tab of the *Operation Parameters* dialog box

Simulate Toolpath

Click the *Simulate Toolpath* button above the graphics area. The *Toolpath Simulation* toolbox appears (like that of Figure 2.29 in Lesson 2). Click the *Run* button to simulate the toolpath. The machining simulation of all five operations will appear in the graphics area, similar to that of Figure 3.16.

Note that the face milling that removes the top layer material of the stock has not been created. Pocket milling and hole drilling operations may not work well in practice without the face milling completed beforehand. We create a face milling operation next.

3.4 Creating a Face Milling Operation

We first tab the CAMWorks feature tree to insert a 2.5 axis feature, and then follow the steps similar to those of Lesson 2 to create a machinable feature for face milling operation.

Manually Create Machinable Feature

Next we learn to manually create a machinable feature for the face milling operation.

From CAMWorks feature tree, right click *Mill Part Setup1* and choose *2.5 Axis Feature* (see Figure 3.17).

Figure 3.16 Material removal simulation

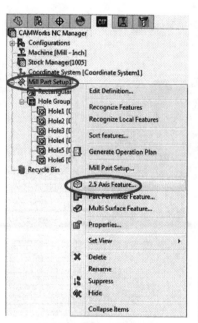

Figure 3.17 Creating a new 2.5 axis feature

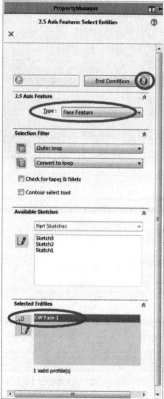

Figure 3.18 The *2.5 axis Feature* dialog box

In the *2.5 Axis Feature* dialog box (Figure 3.18), choose *Face Feature* for *Type*, and pick the top face of the part in the graphics area (see Figure 3.19). The top face is highlighted. The selected face (*CW Face-1*) is now listed under *Selected Entities* in the *2.5 Axis Feature* dialog box (circled in Figure 3.18).

Click the *Next* button ⊕ (circled in Figure 3.18) to define the end condition. Choose *Finish* for *Strategy* (see Figure 3.20), and choose *Upto stock* for the *End condition: Direction 1*. Click the *Reverse direction* button ↯ , if necessary, to make sure *0.25in* appears for dimension, as circled in Figure 3.20. Click the checkmark ✔ to accept the machinable feature.

A *Face Feature1* node is now listed in the CAMWorks feature tree 🔳 in magenta color (see Figure 3.21). Right click *Face Feature1* and choose *Generate Operation Plan*. One new node, *Face Mill1*, is now listed in CAMWorks operation tree 🔳 (Figure 3.22). Right click the node and choose *Generate Toolpath*. A face milling toolpath with a 2in. face mill is generated like that shown in Figure 3.23.

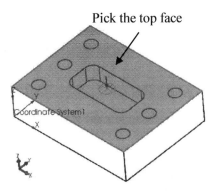

Pick the top face

Figure 3.19 Picking the top face of the part to create a machinable feature for the face milling operation

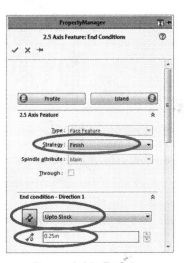

Figure 3.20 Defining end conditions

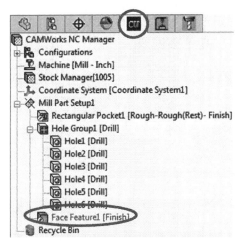

Figure 3.21 A *Face Feature1* machinable feature created

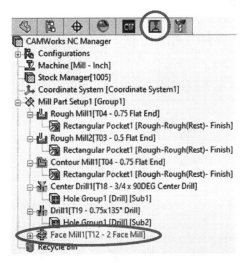

Figure 3.22 A *Face Mill1* operation generated

Choose a Different Cutter

We modify the face milling operation by choosing a 2.5in. face mill (just to learn how to add a cutter to the tool crib) and enter 0.125in. for depth of cut.

Right click *Face Mill1* in the CAMWorks operation tree ⬛ and choose *Edit Definition*. In the *Operation Parameters* dialog box, choose *Tool* tab; the 2in. face mill is shown (see Figure 3.24). Since the 2.5in. face mill is not in the current tool crib, we will need to add it to the crib.

First, we choose the *Tool Crib* tab and click the checkbox in front of the *Filter* button to display only the face mill cutters (see Figure 3.25).

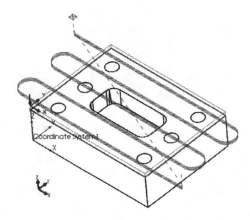

Figure 3.23 The toolpath of *Face Mill1* operation

Click the *Filter* button. In the *Tool Select Filter* dialog box (Figure 3.26), click the checkbox of *Filter by*, select *Face Mill* for *Type*, and then click *OK*. In the *Operation Parameters* dialog box, only one face mill is listed.

Click the *Add* button (see Figure 3.27) to add a face mill tool. In the *Tool Select Filter* dialog box (Figure 3.28), select *Face Mill* for *Tool type*, click the tool in the third row (ID:3) to select the 2.5in. face mill, and then click *OK*.

A 2.5in. face mill is now listed under *Tools* in the *Operation Parameters* dialog box (Figure 3.29). Choose the 2.5in. face mill and click *Select*. Click *Yes* to the question: *Do you want to replace the corresponding holder also?* We have now replaced the cutter with a 2.5in. face mill. We will modify the depth parameters next.

Choose the *Facing* tab of the *Operation Parameters* dialog box (see Figure 3.30). In the *Depth parameters* group, click the percentage button 🔲 to deselect it for both the *First cut amt.* and *Max cut amt.* Enter *0.125in.* for both *First cut amt.* and *Max cut amt.*, as shown in Figure 3.30. Click *OK* to accept the changes. The face milling toolpath will be generated like that shown in Figure 3.5(a) with two rounds of passes, cutting 0.125in. depth in the first round and then the remaining 0.125in. in the next round to finish it up.

Simulate Toolpath

Click the *Simulate Toolpath* button 🔘 above the graphics area. The *Toolpath Simulation* toolbox appears (see Figure 2.29 of Lesson 2). Click the *Run* button ▶ to simulate the toolpath. The material removal simulation of all six operations will appear in the graphics area at the end, similar to that of Figure 3.1.

3.5 Re-Ordering Machining Operations

Now we have created all six operations. However, their order is off. We would like to see these six operations in order as follows: *Face Mill1, Rough Mill1, Rough Mill2, Contour Mill1, Center Drill1*, and *Drill1*, similar to that shown in Figure 3.6. To reorder these operations, you may click and drag them one at a time (or by pressing the shift key to select multiple) and move them up or down in the operation tree.

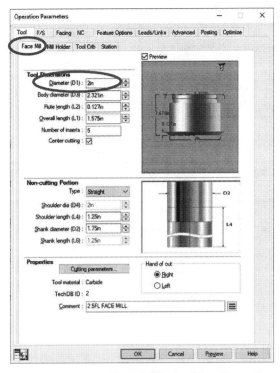

Figure 3.24 The *Face Mill* tab of the *Operation Parameters* dialog box

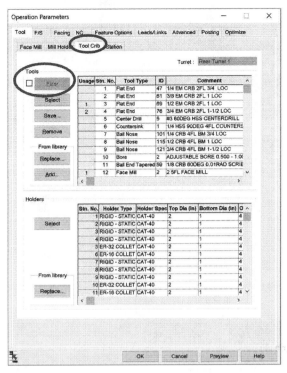

Figure 3.25 Displaying only the face mill cutters using the filter option under the *Tool Crib* tab

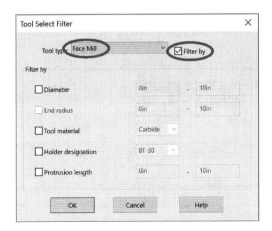

Figure 3.26 The *Tool Select Filter* dialog box

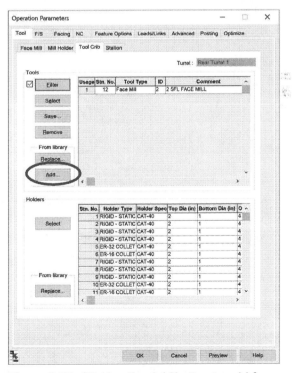

Figure 3.27 Clicking the *Add* button to add face mills under the *Tool Crib* tab

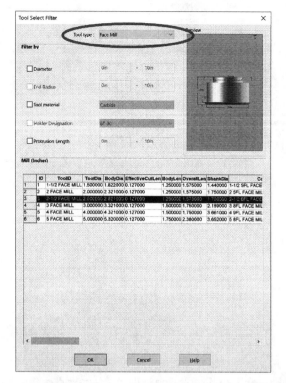

Figure 3.28 Choosing a 2.5in. face mill

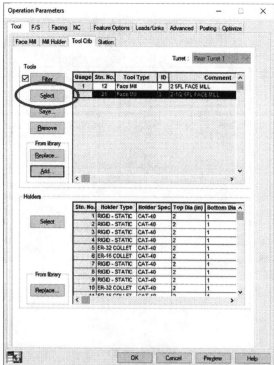

Figure 3.29 Selecting the 2.5in. face mill

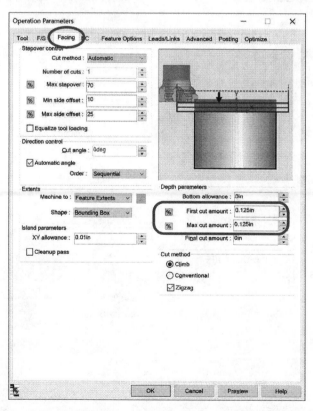

Figure 3.30 Defining depth parameters under the *Facing* tab

In this case, all you have to do is to click and drag *FaceMill1* and release it at above *RoughMill1*.

You may click the *Simulate Toolpath* button ⬡ above the graphics area to review material removal simulation for all six operations combined in a desired order.

3.6   Reviewing Machining Time

You may choose to review machining time for individual operations or the overall time for combined operations.

From CAMWorks operation tree ▣ , right click an operation, for example *Face Mill1*, and choose *Edit Definition*.

In the *Operation Parameters* dialog box (Figure 3.31), choose *Optimize* tab, and look for *Estimated machining time*. The estimated machining time of *Face Mill1* operation is 2.54 minutes.

Note that you may choose *Mill Part Setup1* and right click *Edit Definition* to review the machining time for the combined six operations. Choose *Statistics* tab in the *Part Setup Parameters* dialog box (Figure 3.32). The overall machining time for all six operations is 50.65 minutes. You may calculate the toolpath length by sketching the toolpath (similar to that of Lesson 2) to verify the machining time.

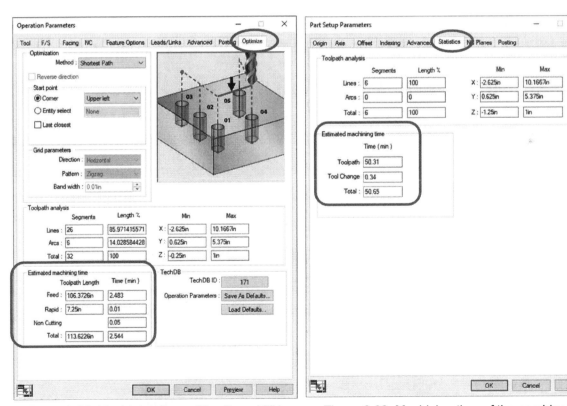

Figure 3.31 Machining time of the *Face Mill1* operation

Figure 3.32 Machining time of the combined six NC operations

```
O0001                    N63 X2.985              N176 G01 X2.5              N328 ( Center Drill1 )
N1 G20                   N64 Y3.615              N177 G00 Z-.15            N329 (3/4 X 90DEG CBT SPOT DRILL)
N2 G91 G28 X0 Y0 Z0      N65 X2.5                N178 Z-.9                 N330 T19 M06
                         N66 G03 X2.385 Y3.5 I0 J-.115   N179 G01 Z-1.25 F1.4381...   N331 S595 M03
N3 ( Face Mill1 )        N67 G01 Y2.5                                      N332 G90 G54 G00 X1. Y5.
N4 (2-1/2 6FL FACE MILL) N68 G03 X2.5 Y2.385 I.115 J0                      N333 G43 Z1. H19 M08
N5 T21 M06               N69 G01 X5.5            N271 ( Contour Mill1 )    N334 G98 G82 Z-.5875 P00 R-.15
N6 S693 M03              ...                     N272 (3/4 EM CRB 2FL 1-1/2 LOC)   F3.3965
N7 G90 G54 G00 X-2.625 Y5.375                    N273 T04 M06              N335 Y3.
N8 G43 Z.1 H21 M08       N152 ( Rough Mill2 )    N274 S1140 M03            N336 Y1.
N9 G01 Z-.125 F5.        N153 (1/2 EM CRB 2FL 1 LOC)   N275 G90 G54 G00 X4.3663 Y3.3273   N337 X7.
N10 G17 X0 F34.3362      N154 T03 M06            N276 G43 Z-.15 H04 M08    N338 Y3.
N11 X8. F45.7816         N155 S1917 M03          N277 G01 Z-.625 F.8556    N339 Y5.
N12 X9.375               N156 G90 G54 G00 X2.5 Y3.74   N278 G41 D24 X4.0905 Y3.603 F2.5669   N340 G80 Z1. M09
N13 G02 Y3.7917 I0 J-.7917   N157 G43 Z-.15 H03 M08   N279 G03 X4.0375 Y3.625 I-.053 J-.053   N341 G91 G28 Z0
N14 G01 X8....           N158 G01 Z-.5 F1.4381   N280 G01 X2.5 F3.4225
                         N159 G03 X2.26 Y3.5 I0 J-.24 F5.7525   N281 G03 X2.375 Y3.5 I0 J-.125   N342 ( Drill1 )
N47 ( Rough Mill1 )      N160 G01 Y3.48          N282 G01 Y2.5             N343 (3/4 SCREW MACH DRILL)
N48 (3/4 EM CRB 2FL 1-1/2 LOC)   N161 G02 X2.52 Y3.74 I.26 J0   N283 G03 X2.5 Y2.375 I.125 J0   N344 T20 M06
N49 T04 M06              N162 G01 X2.5           N284 G01 X5.5             N345 S595 M03
N50 S1140 M03            N163 G00 Z-.15          N285 G03 X5.625 Y2.5 I0 J.125   N346 G90 G54 G00 X1. Y5.
N51 G90 G54 G00 X2.985 Y3.015   N164 Z-.4       N286 G01 Y3.5             N347 G43 Z1. H20 M08
N52 G43 Z-.15 H04 M08    N165 G01 Z-.75 F1.4381  N287 G03 X5.5 Y3.625 I-.125 J0   N348 G98 G83 Z-1.25 Q.1 R-.15 F3.3965
N53 G01 Z-.625 F.8556    N166 G03 X2.26 Y3.5 I0 J-.24 F5.7525   N288 G01 X3.9625   N349 Y3.
N54 Y2.985 F3.4225       N167 G01 Y3.48          N289 G03 X3.9095 Y3.603 I0 J-.075   N350 Y1.
N55 X5.015               N168 G02 X2.52 Y3.74 I.26 J0   N290 G40 G01 X3.6337 Y3.3273   N351 X7.
N56 Y3.015               N169 G01 X2.5           N291 G00 Z-.15            N352 Y3.
N57 X2.985               N170 G00 Z-.15          N292 X4.3663              N353 Y5.
N58 Y3.315               N171 Z-.65              N293 Z-.525               N354 G80 Z1. M09
N59 X2.685               N172 G01 Z-1. F1.4381   N294 G01 Z-.9375 F.8556   N355 G91 G28 Z0
N60 Y2.685               N173 G03 X2.26 Y3.5 I0 J-.24 F5.7525   N295 G41 D24 X4.0905 Y3.603 F2.5669   N356 G28 X0 Y0
N61 X5.315               N174 G01 Y3.48          N296 G03 X4.0375 Y3.625 I-.053 J-.053   N357 M30
N62 Y3.315               N175 G02 X2.52 Y3.74 I.26 J0   N297 G01 X2.5 F3.4225
                                                 ...
```

Figure 3.33 Partial contents of the .txt file (G-code)

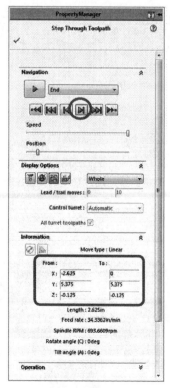

Figure 3.34 The *Step Through Toolpath* dialog box

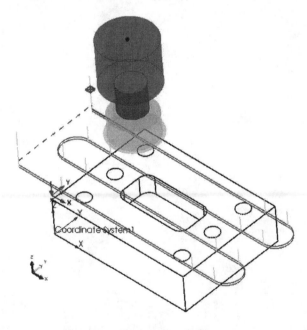

Figure 3.35 The cutter and toolpath displayed in the graphics area

3.7 The Post Process and G-Code

You may click the *Post Process* button ![Post Process] above the graphics area, and follow the same steps learned in Lesson 2 to convert the toolpaths into G-code. Figure 3.33 shows partial contents of the G-code file. We will take a closer look at some of the NC blocks next.

3.8 Stepping Through the Toolpath

We first look at the face milling toolpath by right clicking it (under the CAMWorks operation tree ![icon]) and choosing *Step Through Toolpath*. In the *Step Through Toolpath* dialog box (Figure 3.34), click the *Forward single step* button ![icon] six times to move the cutter to X:0, Y: 5.375, and Z:–0.125 position. The cutter is in contact with the stock, as shown in Figure 3.35. This cutter location takes place after the NC block N10, in which X is 0. Z and Y positions are determined by blocks N9 and N7, respectively (see Figure 3.33). The next block (N11) moves the cutter to the front end of the stock, where X is 8, Y- and Z-coordinates stay the same (Y = 5.375, Z = –0.125). Note that these X-, Y-, and Z-coordinates refer to the part setup origin (again located at the front left corner of the top face of the stock), as they should be.

You may select other operations and step through the toolpaths. You will find the cutter locations (see X, Y, Z shown in the *Step Through Toolpath* dialog box, circled in Figure 3.34) are consistent with those in the G-code.

We have completed this lesson. You may save your model for future reference.

3.9 Exercises

Problem 3.1. Generate machining operations to cut the part shown in Figure 3.36 from a rectangular stock of 4in.×3in.×1.25in. Use cutters and machining parameters chosen by CAMWorks. Please submit the following for grading:

(a) A summary of the NC operations, including number of operations, cutting parameters, and tools selected.
(b) Screen shots of combined NC toolpaths and material removal simulations.

Note that you will have to create two mill part setups to cut the features from the top and bottom of the part, respectively.

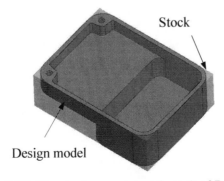

Figure 3.36 The design model and stock of Problem 3.1

Problem 3.2. Use CAMWorks to generate NC operations to cut the part shown in Figure 3.37 from a rectangular block of 4in.×4.25in.×1.55in. (material: steel 1005). Pick your own cutters and choose/enter adequate machining parameters with justifications. Please submit the following for grading:

(a)   A summary of the NC operations, including number of operations, cutting parameters, and tools selected.

(b)   Screen shots of combined NC toolpaths and material removal simulations.

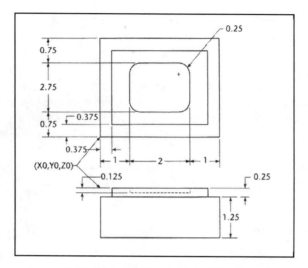

Figure 3.37 The design model of Problem 3.2

Lesson 4: Machining a Freeform Surface

4.1 Overview of the Lesson

In this lesson, we focus on learning virtual machining for cutting a freeform surface (also called contoured or sculptural surface) often seen in mold or die machining. Machining a freeform surface involves rough and finish cuts. Rough cuts remove a major portion of the stock material as fast as possible, often with a larger size tool and hence larger stepover and depth of cut. Multiple rough cuts may be necessary for some applications. Finish cuts polish the freeform surface to meet accuracy and quality requirements.

We will use a 3-axis mill in this lesson to cut a freeform surface. Due to a large curvature variation of the freeform surface in the example employed in this lesson, see Figure 4.1(a), a ball-nose cutter on a 3-axis mill is not able to reach certain areas for an accurate cut, which will become clear later in this lesson. We will revisit this example in Lesson 7: Multiaxis Machining by creating a multiaxis finish cut, in which a 5-axis mill is employed. Figure 4.1(b) shows the material removal simulation for the example of this lesson using a 3-axis mill.

Since a freeform surface is not a standard 2.5 axis feature that AFR is able to extract automatically, we will learn how to manually select surfaces and create a machinable feature. We will then generate operation plan, generate toolpath, and simulate toolpath, similar to what we learned in previous lessons. In this case, CAMWorks generates two NC operations automatically: *Area Clearance* that removes material layer-by-layer normal to the tool axis direction, and *Pattern Project* that polishes the surface by keeping the tool tip in contact with the freeform surface, essentially a surface milling operation.

We will take a closer look at the operations generated by CAMWorks, as well as the options and parameters determined by TechDB™. The *Area Clearance* and *Pattern Project* operations generated by CAMWorks are not quite suitable for this example. For one, the tool chosen for the *Area Clearance* operation is small; therefore, a small stepover and depth of cut are chosen, leading to an extensive machining time.

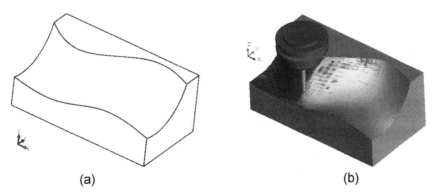

(a)   (b)

Figure 4.1 The freeform surface example, (a) part solid model, and (b) material removal simulation

In general, machining a die with a freeform surface requires a rough cut (often called *volume milling*) that removes material from a raw stock as fast as possible (minimizing machining time) using a larger tool; and therefore, a larger stepover and depth of cut. Since a larger tool and steps are employed, more material remains on the freeform surface uncut at the end of a rough cut. The stock at this stage may not be ready for a finish operation that cuts the stock-in-progress closer to the finished part with desired accuracy and surface quality. Therefore, in practice we often insert a *local milling* in between, in which a smaller tool with smaller stepover and depth of cut are employed to remove the material remaining on the surface of the stock-in-progress after the volume milling. A local milling removes the noticeable amount of material remaining on the surface and prepares the stock-in-progress adequately for the final finish cuts.

In this lesson we learn to implement the three-operation scenario, volume milling, local milling, and surface milling (finish cut), in CAMWorks. We first modify the *Area Clearance* operation by adjusting machining parameter (such as stepover and depth of cut) and selecting a different cutter to remove the material faster. We rename the operation *volume milling*. We then create a local milling (by creating another *Area Clearance*) that removes the remaining material from the first operation using a smaller cutter. Thereafter, we adjust a few parameters and try to improve the surface finish of the surface milling operation (*Pattern Project*) since the default settings do not provide a desired cut. In Lesson 7, we revisit this example by creating a multiaxis surface milling operation for a more desirable surface finish.

4.2   The Freeform Surface Example

The size of the bounding box of the part (filename: *Freeform Surface.SLDPRT*) is 7.5in.×4in.×3in. The solid model is lofted from four parallel sketches, each spaced 2.5in. apart, as shown in Figure 4.2(a). The freeform surface is formed by lofting the circular arcs of respective sketches along the longitudinal direction (or the X-direction of *Coordinate System1*). Individual sketches are created by four straight lines and a circular arc. The front and rear end faces of the loft share the same sketch (that is, Sketch1 and Sketch4 shown in Figure 4.2(a) are identical). Dimensions of individual sketches are shown in Figure 4.2(b), (c), and (d), respectively.

In addition, a coordinate system (*Coordinate System1*) is defined at the front left corner of the bottom face of the part. The unit system chosen is IPS (inch, pound, second). When you open the solid model *Freeform Surface.SLDPRT*, you should see the loft solid feature and a coordinate system listed in the feature tree like that of Figure 4.3.

A stock of retangular block with a size 7.5in.×4in.×3in., made of low carbon alloy steel (1005), as shown in Figure 4.4, is chosen for the machining operations. Note that a part setup origin is defined at the front left vertex on the top face of the stock, which locates the G-code program zero.

As mentioned above, we implement the three-operation scenario for this example. The first operation is a volume milling (*Area Clearance*), which is a rough cut using a 0.75in. hog nose mill with 0.125in. corner radius (T19 in tool crib). The toolpath of the volume milling operation is shown in Figure 4.5(a). The second operation is a local milling (another *Area Clearance*) that continues removing the material remaining from volume milling using a smaller ball nose cutter of diameter 0.5in. (T08); see toolpath in Figure 4.5(b). The third operation is a surface milling (*Pattern Project*) using a 0.25in. tool (T07), serving as a finish cut that produces a machined surface that meets the accuracy and quality requirements; see toolpath in Figure 4.5(c).

You may open the example file with toolpath created (filename: *Freeform Surface with toolpath.SLDPRT*) to preview the toolpath in this example. When you open the file, you should see the three operations listed under the CAMWorks operation tree tab ![icon], as shown in Figure 4.6. You may simulate individual operations by right clicking a node and choosing *Simulate Toolpath*, or simulate the combined operations by clicking the *Simulate Toolpath* button above the graphics area.

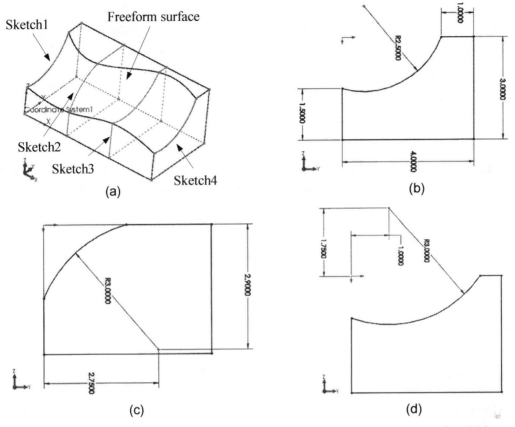

Figure 4.2 Sketches and dimensions of the freeform surface example: (a) the loft solid feature, (b) Sketch 1 (and 4), (3) Sketch 2, and (d) Sketch 3

Figure 4.3 Solid feature tree of the freeform surface example

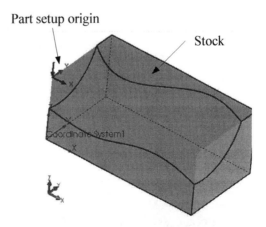

Figure 4.4 Stock with a part setup origin created at the front left vertex on top face

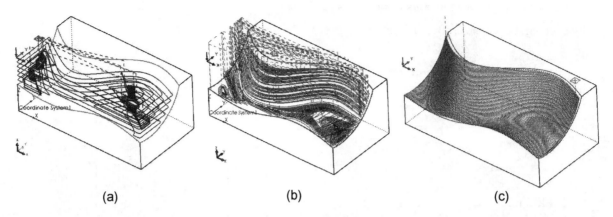

(a)   (b)   (c)

Figure 4.5 Toolpaths of the three NC operations: (a) volume milling (*Area Clearance*), (b) local milling (another *Area Clearance*), and (c) surface milling (*Pattern Project*)

4.3   Using CAMWorks

Open SOLIDWORKS Part

Open the part file (filename: *Freeform Surface.SLDPRT*) downloaded from the publisher's website. This solid model, as shown in Figure 4.2(a), consists of one loft solid feature and a coordinate system. As soon as you open the model, you may want to check the unit system chosen and make sure the IPS system is selected. You may also increase the decimals from the default 2 to 4 digits similar to that of previous lessons.

Select NC Machine

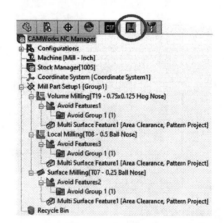

Figure 4.6 The NC operations listed under the CAMWorks operation tree tab

Click the CAMWorks feature tree tab ![icon] and right click *Machine [Mill-inch]* to select *Edit Definition*. Similar to those of Lessons 2 and 3, in the *Machine* dialog box, we select *Mill-inch* under *Machine* tab, choose *Tool Crib2* under *Available tool cribs* of the *Tool Crib* tab, select *M3AXIS-TUTORIAL* under the *Post Processor* tab, and select *Coordinate System1* under *Fixture Coordinate System* of the *Setup* tab.

Create Stock

From CAMWorks feature tree tab ![icon] , right click *Stock Manager* and choose *Edit Definition*. In the *Stock Manager* dialog box, we choose the default stock size (7.5in.×4in.×3in.) and material (*Steel 1005*). The rectangular stock should appear in the graphics window similar to that of Figure 4.4.

We will first select the freeform surface to define a machinable feature. We will let CAMWorks technology database determine machining operations, and then modify them in hopes of generating a better machined surface with reduced machining time. We will relocate the part setup origin to the front left corner at the top face of the stock, as shown in Figure 4.4.

Create a Machinable Feature

Click the CAMWorks feature tree tab ██, right click *Stock Manager*, and choose *Mill Part Setup*, as shown in Figure 4.7.

The *Mill Setup* dialog box appears and the *Front Plane*, perpendicular to the Z-axis, is selected (if not, select *Front Plane*). We choose *Front Plane* to define tool axis of the setup. The *Front Plane* should appear under *Entity* in the dialog box. In the graphics area (see Figure 4.8), a symbol ⊕ with an arrow pointing upward appears. This symbol indicates that the tool axis (or feed direction) must be reversed (pointing downward). Click the *Reverse Selected Entity* button ⬏ under *Entity* to reverse the direction. Make sure that the arrow points in a downward direction. Click the checkmark ✓ to accept the definition. A *Mill Part Setup1* is now listed in the feature tree.

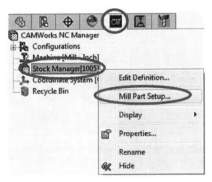

Figure 4.7 Selecting *Mill Part Setup*

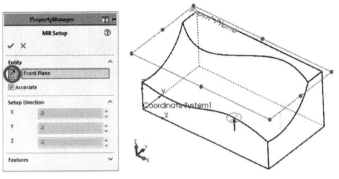

Figure 4.8 Picking the top face for mill setup

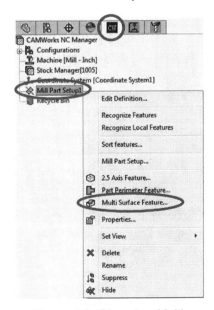

Figure 4.9 Choosing *Multi Surface Feature*

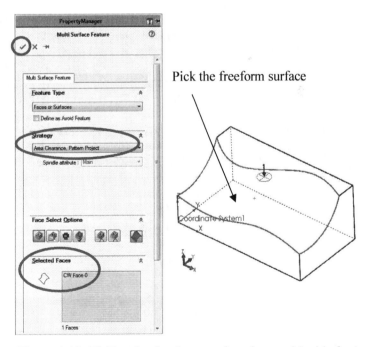

Pick the freeform surface

Figure 4.10 Picking the freeform surface for machinable feature

Now we create a machinable feature. From CAMWorks feature tree ▣, right click *Mill Part Setup1* and choose *Multi Surface Feature* (see Figure 4.9).

In the *Multi Surface Feature* dialog box (Figure 4.10), pick the freeform surface of the part in the graphics area (Figure 4.10); the surface picked is now listed under *Selected Faces*. Leave the default *Area Clearance, Pattern Project* for *Strategy*, and click the checkmark ✔ to accept the surface feature.

In the CAMWorks feature tree ▣, a *Multi Surface Feature1* is added in magenta color under *Mill Part Setup1*, as shown in Figure 4.11.

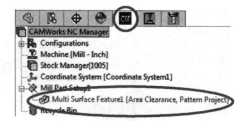

Next we create another multi surface feature and pick the top face of the part to define it as the area to avoid. This is to restrict the toolpath to stay only on the freeform surface.

Right click *Mill Part Setup1* and choose *Multi Surface Feature*.

Figure 4.11

In the *Multi Surface Feature* dialog box, pick the top face of the part in the graphics area (see Figure 4.12), click the *Define as Avoid Feature* box, and click the checkmark ✔ to accept it.

In the CAMWorks feature tree ▣, a *Multi Surface Feature2[Avoid]* is added in magenta color under *Mill Part Setup1*.

Generate Operation Plan and Toolpath

Click the *Generate Operation Plan* button ▤ above the graphics area. Two operations, *Area Clearance1* and *Pattern Project1*, are generated. They are listed in CAMWorks operation tree ▣ (as seen in Figure 4.13). Again they are shown in magenta color, indicating that these operations are not completely defined yet. If you expand an operation (for example, *Area Clearance1*) in the feature tree, an *Avoid Feature1* node is listed. Expand the *Avoid Feature*; *Avoid Group 1 (0)* appears, indicating that the feature to avoid has not been selected (that is why "0" appears in the parentheses). We will accept the operations as they are for now. We will come back shortly to revisit the avoid feature selection.

Note that when you click the *Mill Part Setup1* node (under CAMWorks operation tree ▣), the default part setup origin appears at the front left corner of the bottom face of the stock, coinciding with the coordinate system circled in Figure 4.14, which is less desirable. We will relocate the origin to the front left corner at the top face of the stock.

Redefine the origin by right clicking *Mill Part Setup1* in the CAMWorks operation tree ▣ . In the *Part Setup Parameters* dialog box (see Figure 4.15), choose *Stock vertex* (under the *Origin* tab), and pick the vertex at the front corner of the top face. Click *OK* to accept the change. The part setup origin is now moved to the front left corner of the top face of the stock in the graphics area like that of Figure 4.4.

Click the *Generate Toolpath* button ▥ above the graphics area to create the toolpath. The two operations are turned into black color right after toolpaths are generated.

Toolpaths of the two operations, *Area Clearance1* and *Pattern Project1*, are generated like those in Figure 4.16(a) and Figure 4.17(a), respectively. The toolpath of *Area Clearance1* appear not only removing material above the freeform surface, but cutting the area below the top face of the part (behind the freeform surface). The material removal simulations—shown in Figure 4.16(b)—indicate the cutter plunged into the stock material behind the freeform surface. Furthermore, Figure 4.17(b) reveals that excessive material remained uncut on the freeform surface, leading to undesirable surface finish, after carrying out a material removal simulation for the combined operations.

These toolpaths are certainly not desirable. We need to contain the toolpaths to only machine the freeform surface by selecting the top face of the part as the avoid feature for the *Area Clearance1* operation, and later make an attempt to adjust machining parameters of the *Pattern Project1* to improve the surface finish.

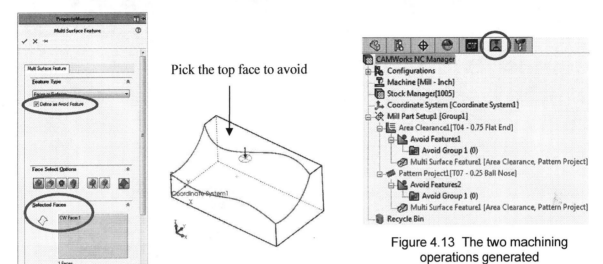

Pick the top face to avoid

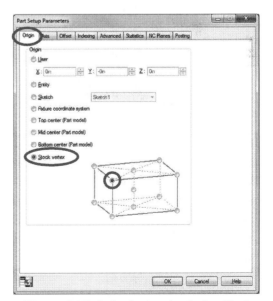

Figure 4.13 The two machining operations generated

Figure 4.12 Picking the top face as a face to avoid

The default part setup origin

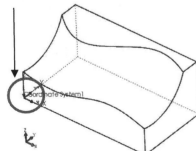

Figure 4.14 The default part setup origin located at the lower left corner of the stock bottom face

Figure 4.15 Selecting vertex in the *Part Setup Parameters* dialog box

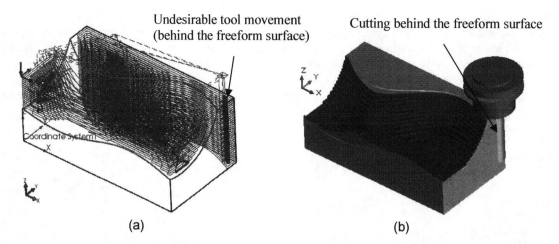

Undesirable tool movement
(behind the freeform surface)

Cutting behind the freeform surface

(a)   (b)

Figure 4.16 The *Area Clearance1* operation, (a) toolpath, and (b) material removal simulation

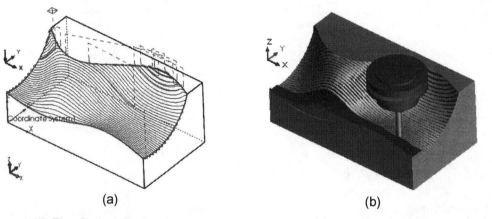

(a)   (b)

Figure 4.17 The *Pattern Project1* operation, (a) toolpath, and (b) material removal simulation
(combined with *Area Clearance* operation)

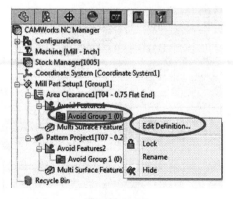

Figure 4.18

Figure 4.19 The *Avoid Features* dialog box

4.4 Selecting Avoid Feature to Correct the Toolpath

Right click *Avoid Group 1(0)* under *Avoid Features1* of *Area Clearance1* and select *Edit Definition* (see Figure 4.18).

Click the small box in front of the *Multi Surface Feature2[Avoid]* in the *Avoid Features* dialog box (Figure 4.19), then click *OK*. The number in the parentheses of *Avoid Group 1* under the feature tree should become *1* (was *0*).

Repeat the same steps for the *Pattern Project1* operation (just to make sure).

Click the *Generate Toolpath* button above the graphics area. The toolpaths will be regenerated like that shown in Figure 4.20. Note that the tool movements of *Area Clearance1* operation behind the freeform surface disappear as desired.

Simulate the toolpaths of combined operations (click the above the graphics area). The material removal *Simulate Toolpath* button simulations shown in Figure 4.21(a) and Figure 4.21(b) indicate that not only is the surface unsmooth but material remained uncut on the freeform surface. How much of the material remained uncut? Let us take a closer look.

In the *Toolpath Simulation* tool box, Figure 4.21(c), select the *Show difference* button under *Display Options*—circled in Figure 4.21(c)—to show a visual comparison of the machined part and the design part in the graphics area. The result is shown in Figure 4.21(d), which indicates the cut is indeed not clean. For example, the area near the middle of the lower edge of the freeform surface circled in Figure 4.21(d) shows dark blue color indicating an amount of more than 0.0157in. undercut material, as depicted in the color spectrum to the left. Other areas, such as those near the top edge, show a similar issue.

How long does the entire machining operation take? The machining times are estimated as 321 and 8.6 minutes, respectively, for *Area Clearance1* and *Pattern Project1* operations, based on the feedrates determined by the TechDB™. Can the machining time of the *Area Clearance1* operation be reduced? Is the toolpath of *Pattern Project1* fine enough to produce a satisfactory finished surface? Can we modify the toolpath of the rough cut (*Area Clearance1*) for a reduced machining time and the toolpath of the finish cut (*Pattern Project1*) for a better surface finish and accuracy? Furthermore, we did find the areas that show deviation between the finished part and the design part in Figure 4.21(d). Can this deviation be reduced or completely eliminated?

4.5 Modifying the Area Clearance Toolpath

How do we modify the toolpath? Where to find and modify machining parameters that adjust the toolpath, such as depth of cut, stepover, and cut pattern? First we take a closer look at the first operation, *Area Clearance1*. We would like to use a different cutter and larger stepover and depth of cut to reduce the machining time.

In CAMWorks operation tree , right click *Area Clearance1* and choose *Edit Definition*. In the *Operation Parameters* dialog box, Figure 4.22(a), T76 (tool ID designated in TechDB™) has been selected (0.75in. flat-end mill) under the *Tool* tab. Ideally, we would like to use a large-size cutter for this operation. However, there is no larger cutter that is suitable for this operation in the tool library. Hence, we will use a 0.75in. (same outer diameter) hog-nose cutter with a corner radius 0.125in. for this operation.

In the *Operation Parameters* dialog box, choose the *Tool Crib* tab. Similar to that of Lesson 3, we use the *Filter* option—see Figure 4.22(b) and Figure 4.22(c)—to display hog nose cutters only. Since there is no hog nose cutter in the tool crib, we will have to add one from the tool library we intend to use.

Click *Add* under the *Tool Crib* tab; see Figure 4.22(b). In the *Tool Select Filter* dialog box, select tool ID 170 (3/4 CRB 4FL HGN .125R 1-1/2 LOC) and click *OK*; see Figure 4.22(d). The tool will be listed under the *Tool Crib* tab; see Figure 4.22(e). Click *Select*. Click *Yes* to the question: *Do you want to replace the corresponding holder also?* Choose *Mill Tool* tab to review details of the hog nose cutter selected; see Figure 4.22(f).

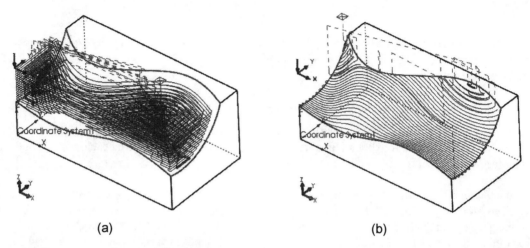

(a)   (b)

Figure 4.20 Toolpaths with avoid features selected, (a) *Area Clearance1*, and (b) *Pattern Project1*

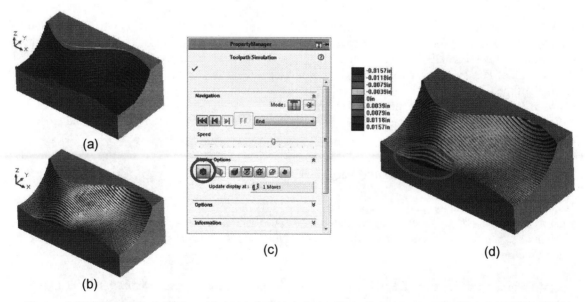

(a)

(b)

(c)

(d)

Figure 4.21 The material removal simulation, (a) *Area Clearance1* operation only (b) combined operations, (c) the *Show Difference* button of the *Toolpath Simulation* tool box, and (d) differences between the machined part and the design part shown with a color spectrum

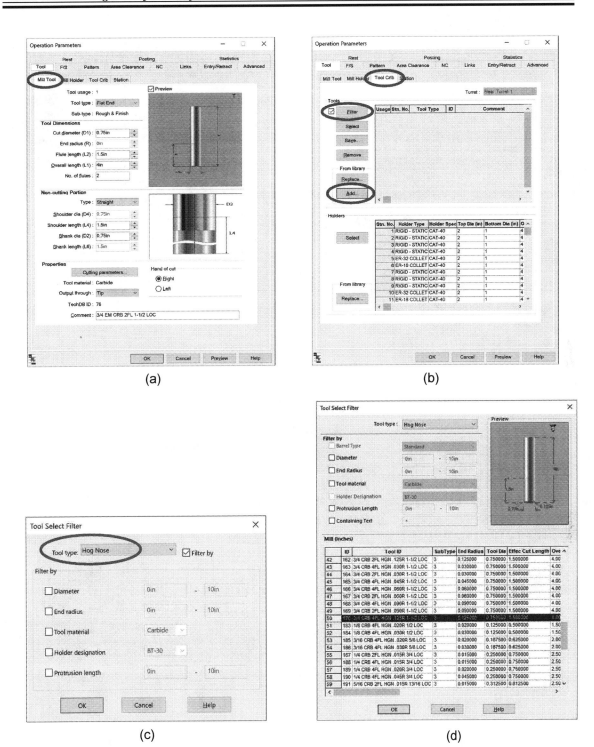

Figure 4.22 Replacing the tool with a 0.75in. hog nose, (a) 0.75in. flat-end mill selected currently, (b) showing no hog nose cutter in the tool crib using the filter option, (c) the tool select filter, (d) adding the 0.75in. hog nose cutter (ID: 170), (e) the hog nose cutter listed, and (f) the 0.75in. hog nose cutter selected

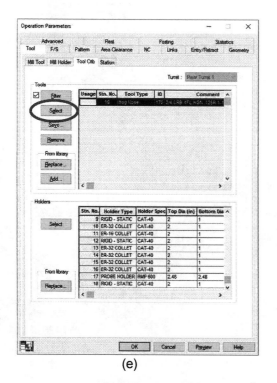

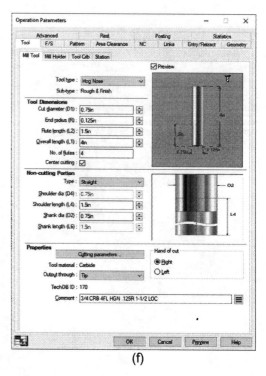

(e)   (f)

Figure 4.22 Replacing the tool with a 0.75in. hog nose (cont'd)

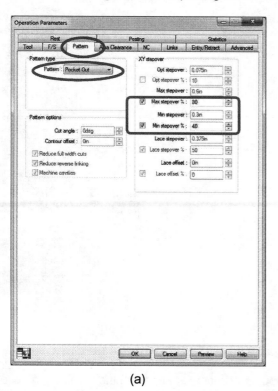

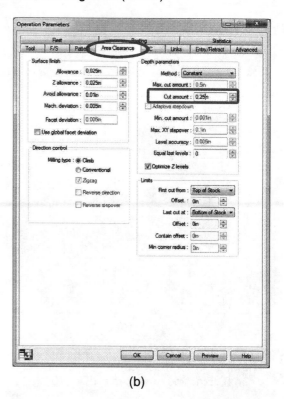

(a)   (b)

Figure 4.23 Modifying machining parameters, (a) edit parameter/option under the *Pattern* tab, (b) change *Cut amount* under the *Area Clearance* tab, and (c) show machining time in the *Statistics* tab

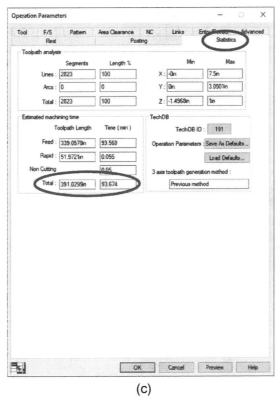

(c)

Figure 4.23 Modifying machining parameters (cont'd)

We have now replaced the tool with a modified 0.75in. hog nose cutter with 0.125in. corner radius. We will modify the machining parameters next.

Choose the *Pattern* tab of the *Operation Parameters* dialog box, select *Pocket Out* for *Pattern*, and enter 80% and 40% for *Max.* and *Min. stepover*, respectively, as shown in Figure 4.23(a).

Choose the *Area Clearance* tab of the *Operation Parameters* dialog box; see Figure 4.23(b). In the *Depth* parameters group, enter *0.25in.* for *Cut amount*, as shown in Figure 4.23(b). Note that the cut amount 0.25in. is 1/3 of the cutter diameter, which is considered adequate in practice in general. Click *OK* to accept the changes and click *Yes* to the warning message: *Operation parameters have changed, toolpaths need to be recalculated. Regenerate toolpaths now?*

The area clearance toolpath will be regenerated like that shown in Figure 4.5(a). The machining time is now 93.7 minutes (click the *Statistics* tab to see the machining time estimated), as shown in Figure 4.23(c). Although the machining time is reduced, is the operation acceptable and is the stock-in-progress ready for the next operation, *Pattern Project1*?

Right click *Area Clearance1* and choose *Rename*. Change the name of the operation to *Volume Milling*.

4.6   Reviewing the Machined Part Quality

Now let us take a closer look at the stock-in-progress after carrying out the *Volume Milling* operation. We will learn to use a few more buttons in the *Simulate Toolpath* toolbox, in particular the *Section view* button to create section views.

Right click *Volume Milling* and select *Simulate Toolpath*. The *Toolpath Simulation* toolbox appears; see Figure 2.29 of Lesson 2. Click the *Run* button ▶ to simulate the toolpath. The material removal simulation of the *Volume Milling* operation will appear in the graphics area at the end, similar to that of Figure 4.24(a).

Click the *Stock Display* button 🔲 and choose *Wireframe Display* to display the stock in wireframe—see Figure 4.24(b)—which shows a better view in terms of the geometric shape of the machined surface in progress.

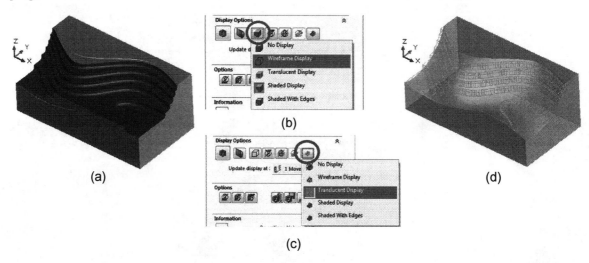

(a)   (b)

(c)   (d)

Figure 4.24 Options of the *Simulate Toolpath* tool box, (a) material removal simulation, (b) options to display stock in wireframe, (c) options to display target part in translucent, and (d) tool display turned off and target part in translucent display

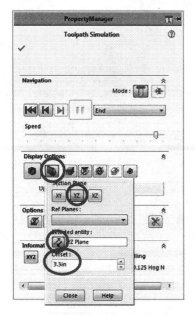

Figure 4.25 Defining a section view in the *Toolpath Simulation* toolbox

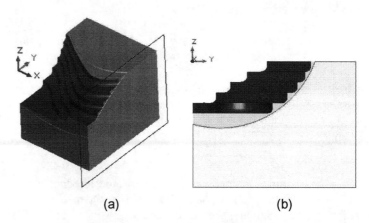

(a)   (b)

Figure 4.26 Section views of YZ plane, (a) current work view, and (b) right end view

You may click the *Tool Display* button 🔳 and choose *No Display* to turn off the tool display, and click the *Target Part Display* button 🔳 and choose *Translucent Display*—see Figure 4.24(c)—showing the design (or target) part overlapping with the stock-in-progress; see Figure 4.24(d). Boundary edges of the design part shown in light green color clearly differentiate the significant amount of material that remained uncut after the *Volume Milling* operation.

You may create a section view to review the quality of the machined part at specific sections normal to a direction of a selected plane. Click the *Section view* button 🔳 from the *Toolpath Simulation* toolbox (Figure 4.25), choose *YZ* for *Plane*, click the *Reverse direction* button 🔳, and increase the *Offset* to *3.3in.* (see Figure 4.25). A section view appears like that of Figure 4.26(a), which shows the difference between the target and the machined parts at the section.

You may rotate the view to see the section; for example, viewing it from the right end; see Figure 4.26(b). It is apparent that there is a significant amount of material uncut. Next, we create a local milling operation to remove the uncut material with a smaller tool and reduced stepover and depth of cut.

4.7   Adding a Local Milling Operation

We will add a local milling operation by creating another area clearance operation to continue cutting the material remaining from the *Volume Milling* operation with a ball nose cutter of 0.5in. diameter. We enter *0.2in* for both stepover and depth of cut.

Under the CAMWorks operation tree 🔳, right click *Mill Part Setup1* and choose *3 Axis Mill Operations > Area Clearance*.

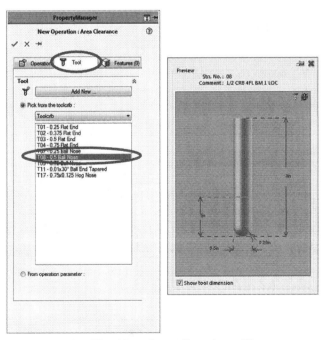

Figure 4.27 The *New Operation: Area Clearance* dialog box

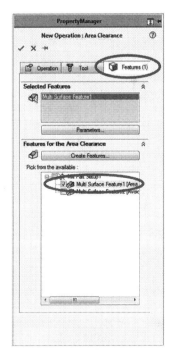

Figure 4.28 The *Features* tab of the *New Operation: Area Clearance* dialog box

The *New Operation: Area Clearance* dialog box (Figure 4.27) appears. Under the *Tool* tab, select *T08 – 0.5 Ball Nose*; a sketch of the tool with dimensions appears to the right (see Figure 4.27).

Click the *Features* tab, select *Multi Surface Feature1 [Area Clearance, Pattern Project]*, and click the checkmark ✔ to accept the new operation (see Figure 4.28).

A new operation, *Area Clearance2*, is now listed in the CAMWorks operation tree tab ![icon] (see Figure 4.29) and the *Operation Parameters* dialog box appears.

Choose the *Pattern* tab of the *Operation Parameters* dialog box, select *Pocket Out* for *Pattern*, enter *20* for both *Max stepover%* and *Min stepover%*, as shown in Figure 4.30(a).

Choose the *Area Clearance* tab of the *Operation Parameters* dialog box. In the *Depth* parameters group, enter *0.1in* for *Cut amount*, as shown in Figure 4.30(b).

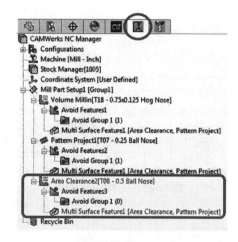

Choose the *Rest* tab of the *Operation Parameters* dialog box, as shown in Figure 4.31(a). Choose *From WIP* for *Method* under *Rest Machining*, and click the selection button ▭ to select *Volume Milling* for *Compute WIP from operations* in the *Operations for WIP* dialog box; see Figure 4.31(b).

This is how we link the local milling to the previous operation(s). WIP stands for work in progress. Click *OK* to accept the changes.

Expand the newly created operation (*Area Clearance2*) to see if the number in the parentheses in *Avoid Group 1* is zero. If not, right click *Avoid Group 1* to add *Multi Surface Feature2* as the avoid feature (similar to Figure 4.19). The number in the parentheses in *Avoid Group 1* should be *1*.

Figure 4.29 *Area Clearance2* added to the feature tree

Right click *Area Clearance2* and choose *Generate Toolpath* to generate a toolpath like that shown in Figure 4.5(b).

Right click again *Area Clearance2* and choose *Rename*. Change the name of the operation to *Local Milling*. The machining time for the *Local Milling* operation is 32.1 min., calculated using the default feedrate chosen by TechDB™.

Simulate the toolpaths for combined *Volume Milling* and *Local Milling* by, for example, suppressing *Pattern Project1* (right click *Pattern Project1* and choose *Suppress*) and clicking the *Simulate Toolpath* button ⬚ above the graphics area. Click the *Run* button ▶ to simulate the toolpath. The material removal simulation of the two operations combined appear in the graphics area, similar to that of Figure 4.32.

By visually inspecting the material removal simulations of the combined *Volume Milling* and *Local Milling* (Figure 4.32) with that of *Area Clearance1* generated by CAMWorks with machining parameters chosen by TechDB™ shown in Figure 4.33, the quality of the machined part from the combined operations is better judging by the material that remained uncut, as can also be seen at the section views.

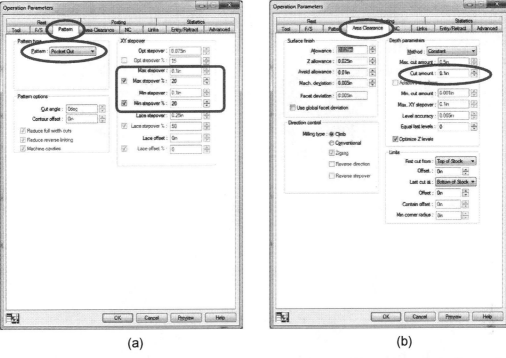

(a) (b)

Figure 4.30 Defining the new operation: *Area Clearance2*, (a) editing parameter/option under the *Pattern* tab, and (b) changing cut amount under the *Area Clearance* tab

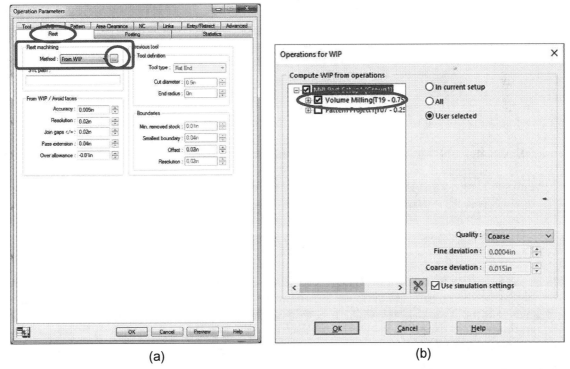

(a) (b)

Figure 4.31 Linking local milling to a previous operation, (a) choosing method under the *Rest* tab, and (b) picking *Volume Milling* in the *Operations for WIP* dialog box

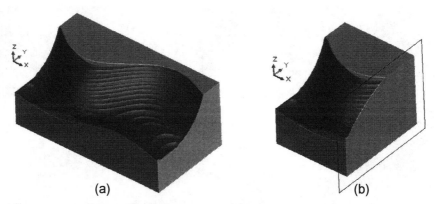

Figure 4.32 Material removal simulation for *Volume Milling* and *Local Milling*, (a) current work view, and (b) section view

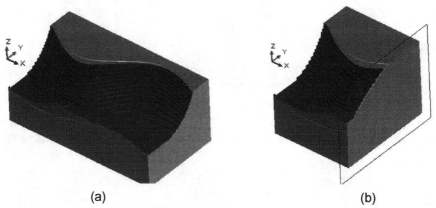

Figure 4.33 Material removal simulation for *Area Clearance1*, (a) current work view, and (b) section view

The machining time of the combined operations is 126 min. (93.7 + 32.1), less than that of *Area Clearance1* (321 min.), all calculated using default feedrates chosen by TechDB™.

4.8   Reviewing and Modifying the Pattern Project Operation

As shown in Figure 4.21(c), the machined surface after the *Pattern Project1* operation can be improved by entering a smaller stepover, for example, reducing the stepover from 0.1 to 0.05in. This can be done by right clicking *Pattern Project1* node under CAMWorks operation tree ▣, and selecting *Edit Definition*. In the *Operation Parameters* dialog box (see Figure 4.34), choose *Pattern* tab, leave *Slice* pattern type and enter a stepover *0.05in.*

Regenerate toolpath for the *Pattern Project* operation, as seen in Figure 4.35(a). The toolpath seems to be fine except that the toolpath near the lower edge—circled in Figure 4.35(a)—may leave a noticeable amount of material uncut. A similar observation is noted near the top edge of the freeform surface. Note that the cutting time estimated for the revised *Pattern Project* operation is 11.9 minutes, slightly increased from 8.5 minutes before revision.

We change the name of operation *Pattern Project1* to *Surface Milling*.

Now, we simulate the toolpaths for all three operations combined: *Volume Milling*, *Local Milling*, and *Surface Milling*.

Click the *Simulate Toolpath* button ⬡ Simulate Toolpath above the graphics area. Click the *Run* button ▶ to simulate the operations. The material removal simulation of the combined three operations appears in the graphics area, similar to that of Figure 4.35(b) and Figure 4.35(c) (section view). By visual inspection, the quality of the machined surface is in general very good, except that a noticeable amount of material uncut near the lower edge of the freeform surface—see the area circled in Figure 4.35(b)—is apparent, as predicted. Same is true near the top edge, but less severe. The surface quality is much better comparing Figure 4.35(b) with Figure 4.21(d).

How do we adjust the toolpath of the *Surface Milling* operation to further improve the accuracy of the machined part? We will explore more on altering the options and parameters of the operation to see if an improved operation is possible. We will explore a different pattern type—more specifically, the flowline pattern, and alter the direction of the toolpath to see if the uncut areas can be minimized.

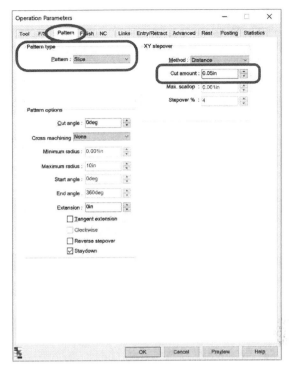

Figure 4.34 The *Pattern* tab of the *Operation Parameters* dialog box

Under CAMWorks operation tree ⬚ , right click *Surface Milling* and choose *Edit Definition*. In the *Operation Parameters* dialog box, click the *Pattern* tab, as shown in Figure 4.36(a), to review the pattern type and stepover. Change the pattern to *Flowline* and change the stepover to 0.05 in. (*Cut Amount* under *XY stepover*), which is small.

Click *Curve 1*; the *Curve Wizard*, as shown in Figure 4.36(b), appears. Select *Single Edge*, pick the top edge curve of the freeform surface in the graphics area; see Figure 4.36(c). A dot appears at the front end of the edge curve; see Figure 4.36(c). This dot is considered the end point of the edge curve since we expect the toolpath to start on the top edge of the freeform surface. We select the *End of Curve* under *Start point*. As soon as the *End of Curve* under *Start point* is selected, the point at the rear end of the curve (*Curve 1*) is highlighted, indicating the point is the start point of the edge curve. Click the checkmark ✔ to accept the selection.

Click *Curve 2*, and pick the lower edge curve of the freeform surface in the graphics area. Notice that the point at the rear end of the curve (*Curve 2*) is highlighted, indicating the point is already assigned as the start point of the curve. No change is needed. This ensures that the start points of both curves are consistent leading to a valid toolpath. Click the checkmark ✔ to accept the selection.

Leave *Along* for *Cut*, and click the *Preview* button to show the toolpath like that of Figure 4.37(a). Note that the toolpath still leaves a noticeable gap at the lower edge (and top edge) as circled in Figure 4.37(a). We accept the toolpath for the time being. Click the close button ⊠ on the top right corner of the *Operation*

Parameters dialog box—circled in Figure 4.36(a)—and click *OK* to accept the changes and regenerate the toolpath.

Simulate the toolpaths for the combined three operations, as shown in Figure 4.37(b), and use section view to review the machined surface; see Figure 4.37(c).

The material removal simulation indicates that there is still a noticeable amount of material remaining on the freeform surface, especially near the lower edge (and some near the upper edge) in the area circled in Figure 4.37(b).

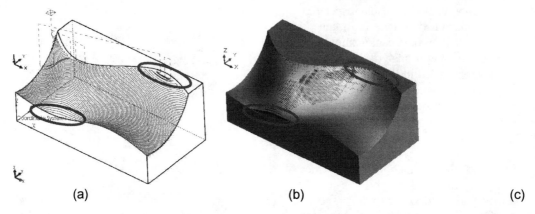

(a)   (b)   (c)

Figure 4.35 Toolpath and material removal simulation, (a) toolpath of *Surface Milling* (choosing *Slice* for *Pattern*), (b) material removal simulation of the combined three operations, and (c) section view at offset 3.3in. in YZ plane

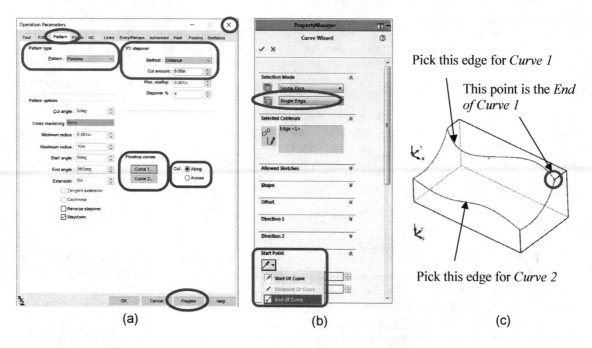

(a)   (b)   (c)

Figure 4.36 Defining a *Flowline* type operation, (a) choosing *Flowline* for *Pattern* type, (b) the curve wizard showing single edge, curve picked, and options of determining start point, and (c) picking curves

Now we alter the toolpath direction by choosing *Across* for *Cut* in the *Pattern* tab of the *Operation Parameters* dialog box; see Figure 4.36(a). The toolpath still leaves a noticeable gap near the lower edge (and less at close to the top edge) as circled in Figure 4.38(a). The material removal simulation indicates slightly better results at the top edge, as circled in Figure 4.38(b). There is still material remaining near the lower edge, as shown in Figure 4.38(b) and a section view in Figure 4.38(c).

With a 0.25in. ball nose cutter and 3-axis mill, the curvature of the convex area near the lower edge makes the surface too steep for the tool to reach, leading to the noticeable uncut material. This problem can be properly addressed using a 5-axis mill. We will learn to generate operations using 5-axis mill in Lesson 7, and we will revisit this example at the time.

We have completed this exercise. You may save your model for future reference. Please keep the model file since we will need it in Lesson 7.

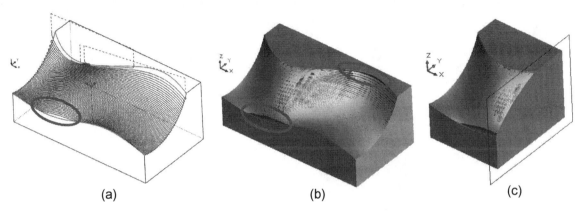

(a)   (b)   (c)

Figure 4.37 Toolpath and material removal simulation, (a) toolpath of the *Surface Milling* operation (choosing *Along* for *Cut*), (b) material removal simulation of combined operations, and (c) section view at offset 3.3in. in YZ plane

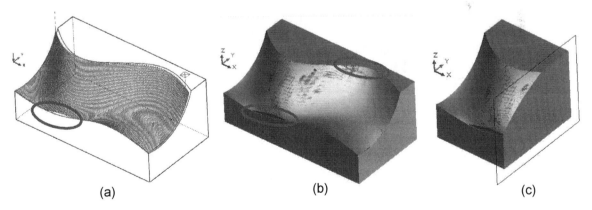

(a)   (b)   (c)

Figure 4.38 Toolpath and material removal simulation, (a) toolpath of the *Surface Milling* operation (choosing *Across* for *Cut*), (b) material removal simulation of combined operations, and (c) section view at offset 3.3in. in YZ plane

4.9 Exercises

Problem 4.1. Follow the same steps discussed in this lesson to machine a part shown in Figure 4.39 using a stock of 12in.×5.5in.×3.5in. Note that at least three operations using a 3-axis mill must be included:

volume milling, local milling, and surface milling. Report the tools and machining options selected for individual operations. Also report machining times of individual and combined operations. Would machining operations of 3-axis mill give you a satisfactory machined part? Do you notice uncut areas like those in, for example, Figure 4.37?

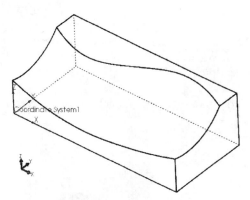

Figure 4.39 Part of Problem 4.1

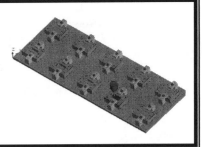

Lesson 5: Multipart Machining

5.1 Overview of the Lesson

So far, we have discussed virtual machining for cutting a single part that is created as a part solid model in SOLIDWORKS. In this lesson, we will move one step further, in which we focus on creating machining operations for a set of identical parts in an assembly solid model created in SOLIDWORKS. Machining multiple parts in a single setup is a common practice at shop floor.

The individual part to be machined is identical to that in Lesson 3, which involves pocket milling and hole drilling using a 3-axis mill. The face milling operation is not included in this lesson to simplify the fixture design that clamps the stock to a jig table. In SOLIDWORKS assembly, the part (more precisely, the stock) is assembled to a jig table by using two fixtures (one on each side). Each fixture consists of a clamp, a riser, and a threaded bolt. A total of ten parts arranged in two rows are to be machined, as shown in Figure 5.1.

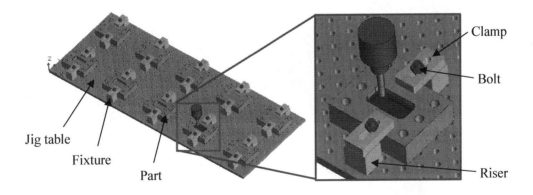

Figure 5.1 The material removal simulation of the multipart machining example

In this lesson, we learn the steps to create instances of the part in CAMWorks, define stocks for individual instances, extract machinable features, generate toolpath, and select components in the assembly (including the jig table and fixtures) for the tools to avoid. In addition, we will take a closer look at the G-code generated by CAMWorks for machining multiple parts in an assembly.

5.2 The Multipart Machining Example

The size of the bounding box of the part (filename: *2 point 5 axis features.SLDPRT*) is 8in.×6in.×2in. There is a center pocket and six holes, three on each side, as shown in Figure 5.2(a). As discussed in Lesson 3, these are 2.5 axis features that can be extracted automatically as machinable features by using the automatic feature recognition (AFR) capability. The stock size is identical to that of the bounding box.

As mentioned earlier, the face milling operation discussed in Lesson 3 is excluded in this lesson. The pocket milling operations, including two rough mills (Rough Mill1 and Rough Mill2) and a contour mill—toolpaths shown in Figure 5.2(b)—and hole drilling operations—center drill and drill shown in Figure 5.2(c)—are identical to those of Lesson 3.

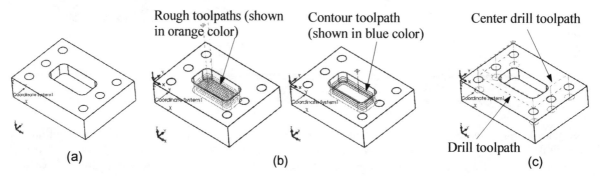

Figure 5.2 The *2 and 5 axis features* example: (a) part solid model, (b) pocket milling toolpaths: two rough and one contour, and (c) hole drilling toolpaths: center drill and drill

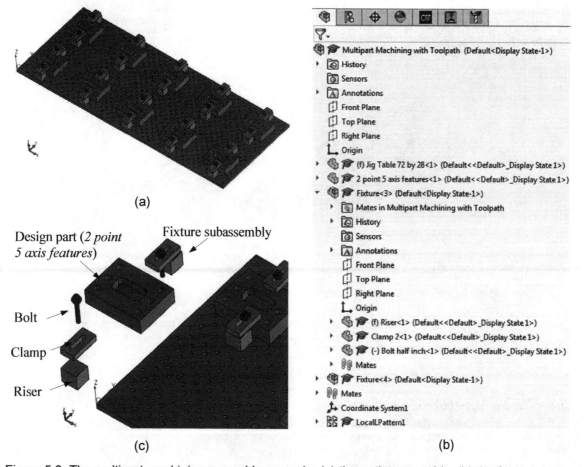

Figure 5.3 The multipart machining assembly example: (a) the entire assembly, (b) the feature tree, and (c) explode and zoom-in view

The stock is clamped on the jig table by two fixtures, one on each side. The jig table is 72in.×28in.×0.75in., with threaded holes of 0.5in. in diameter and 0.5in. in depth. The holes are 1.5in. apart every other row and column; that is, along the X and Y directions, respectively, of the coordinate system, *Coordinate System1*. This coordinate system is located at the front left corner of the jig table, as shown in Figure 5.3(a) and Figure 5.3(c).

The solid models *2 and 5 axis features.SLDPRT* are the only parts (as in ten instances) that will be machined. The fixtures are created as subassemblies; each consists of three parts, riser, clamp, and a bolt, as seen in Figure 5.3(c). In this lesson, we choose *Coordinate System1* for both fixture coordinate system and part setup origin. As a result, the G-code generated refers to this coordinate system. The design part (*2 and 5 axis features.SLDPRT*) with the two fixtures are patterned to create an additional nine instances in two rows as a linear pattern feature (*LocalPattern1* under the FeatureManager design tree). The unit system chosen is IPS (inch, pound, second). When you open the assembly model *Multipart Machining.SLDASM*, you should see 2 parts and 2 subassemblies, a coordinate system, and a pattern feature, listed in the feature tree like that of Figure 5.3(b).

You may open the example file with toolpath created (filename: *Multipart Machining with Toolpath.SLDASM*) to preview toolpath created for this example. When you open the file, you may expand the *Part Manager* node, expand *2 point 5 axis features.SLDPRT*, and then the *Instances* to see the ten instances of the part to be machined (*2 and 5 axis features<1>* to *<10>*), as shown in Figure 5.4.

Also, you should see the five operations listed under *Setup1* of the CAMWorks operation tree tab (see Figure 5.4).

You may click the *Simulate Toolpath* button above the graphics area to preview the machining simulation.

5.3   Using CAMWorks

Open SOLIDWORKS Assembly

Open the assembly model (filename: *Multipart Machining.SLDASM*) downloaded from the publisher's website. This assembly model, as shown in Figure 5.3, consists of four components (jig table, the part: *2 and 5 axis features*, and two fixtures), nine mates, a coordinate system, and a linear pattern. Again, as soon as you open the assembly model, you may want to check the unit system and make sure the IPS system is selected. You may also increase the decimals from the default 2 to 4 digits similar to that of previous lessons.

Select NC Machine

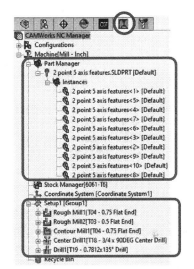

Figure 5.4 The instances and NC operations listed under the CAMWorks operation tree tab

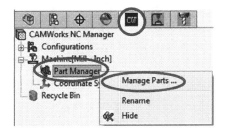

Figure 5.5 Right clicking *Part Manager* and selecting *Manage Parts*

Click the CAMWorks feature tree tab and right click *Mill-inch* to select *Edit Definition*. Similar to those of previous lessons, in the *Machine* dialog box, we select *Mill-inch* under *Machine* tab, choose *Tool*

Crib 2 under *Available tool cribs* of the *Tool Crib* tab, select *M3AXIS-TUTORIAL* under the *Post Processor* tab, and select *Coordinate System1* under *Fixture Coordinate System* of the *Setup* tab.

Manage Part

Since we are dealing with an assembly with multiple components, we have to identify which part or parts to cut. Under the CAMWorks feature tree tab ▢, right click *Part Manager* and select *Manage Parts* (see Figure 5.5). The *Manage Parts* dialog box appears (see Figure 5.6).

Pick the part (*2 point 5 axis features*) in the graphics area or select the node, *2 point 5 axis features*, under the FeatureManager design tree tab ▢. The part is now listed under *Selected Parts* in the *Manage Parts* dialog box (circled in Figure 5.6). Click the part and click *Add All Instances* button to bring in all instances, and then click *OK* to accept the part definition.

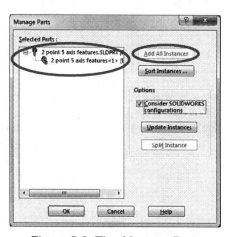

Figure 5.6 The *Manage Parts* dialog box

Click the CAMWorks feature tree tab ▢, expand the *Part Manager* node to see the part (*2 point 5 axis features.SLDPRT*) and its instances like that of Figure 5.7. Also, a *Stock Manager* node is added to the feature tree, as circled in Figure 5.7.

Create Stock for Instances

From the CAMWorks feature tree ▢, right click *Stock Manager* and choose *Edit Definition*. In the *Stock Manager* dialog box (Figure 5.8), we use the default stock size (8in.×6in.×2in.) and choose material: *Aluminum 6061-T6*, and click the *Apply Current Definitions to All Parts* button ▢ (circled in Figure 5.8). Then click the checkmark ✔ to accept the stock definition.

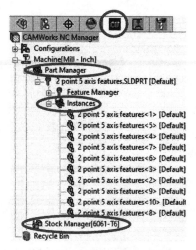

Figure 5.7 Entities listed under the CAMWorks feature tree tab

Figure 5.8 The *Stock Manager* dialog box

Extract Machinable Features

Click the *Extract Machinable Features* button ![btn] above the graphics area. A *Setup1* node is created with two machinable features extracted: *Rectangular Pocket* and *Hole Group* (with six holes), all listed in CAMWorks feature tree tab ![tab] (Figure 5.9). Both machinable features are shown in magenta color.

In the graphics area (see Figure 5.10), a symbol ![sym] with arrow pointing downward appears at the front left corner of the jig table coinciding with *Coordinate System1* (not shown in Figure 5.10). This symbol indicates that the tool axis is chosen correctly (pointing downward). You may need to move the mouse point over *Setup1* in the feature tree to show the part setup origin symbol in the graphics area (circled in Figure 5.10). This corner point also serves as the part setup origin, which locates G-code program zero.

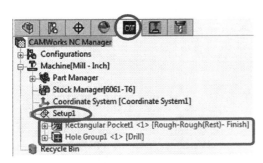

Figure 5.10 The two machinable features extracted

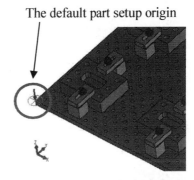

The default part setup origin

Figure 5.9 The default part setup origin coinciding with *Coordinate System1*

Generate Operation Plan and Toolpath

Click the *Generate Operation Plan* button ![btn] above the graphics area. Five operations, Rough Mill1 (cutting the pocket), Rough Mill2 (cutting the corner fillets of the pocket), Contour Mill1 (cutting the boundary faces of the pocket), Center Drill1, and Drill1, are generated. They are listed in CAMWorks operation tree (Figure 5.11). Again they are shown in magenta color. Click the *Generate Toolpath* button ![btn] above the graphics area to create the toolpath. The five operations are turned into black color after toolpaths are generated.

Figure 5.11 The five NC operations generated

Part Setup Origin

As pointed out earlier, if you click *Setup1* under CAMWorks operation tree ![tab], the part setup origin appears at the front left corner of the jig table, coinciding with the fixture coordinate system. Next, we review the options that define the setup origin for machining multiple parts in assembly.

Click the CAMWorks operation tree tab ![tab], and right click *Setup1* to choose *Edit Definition*. In the *Setup Parameters* dialog box, click the *Origin* tab (Figure 5.12). Choose *Setup origin* under *Output origin*, choose *Fixture coordinate system* under *Setup origin*, as circled in Figure 5.12, and leave *0* for *X:*, *Y:*, and *Z:* coordinates. This implies that the G-code will be output referring back to the SOLIDWORKS assembly coordinate system, which is *Coordinate System1*.

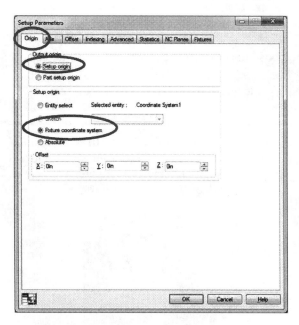

Figure 5.12 The *Origin* tab in the *Setup
Parameters* dialog box

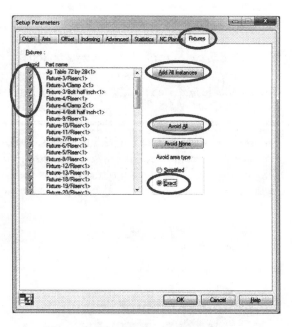

Figure 5.13 The *Fixtures* tab in the
Setup Parameters dialog box

We review the G-code output by CAMWorks in Section 5.5, based on the selections. We will explore other options in the exercise problems at the end of the lesson.

Fixtures and Components to Avoid

We will select the jig table and the two fixtures to avoid in generating the toolpath.

In the *Setup Parameters* dialog box, click the *Fixtures* tab (Figure 5.13). Choose the components to avoid, including the jig table and riser, clamp, and bolt of the two fixture subassemblies, from the FeatureManager design tree tab ![icon] (see Figure 5.14).

In the *Setup Parameters* dialog box, choose all components listed (by clicking the small checkboxes in front, as circled in Figure 5.13), and click *Add All Instances*, *Avoid All*, and *Exact*, then click *OK*. Please make sure you select *Exact*. If not, you may see undesirable toolpath, for example, in Rough Mill2 operation. You may also need to choose one part at a time (by clicking the small box) and click *Add All Instances* to make sure all instances are included.

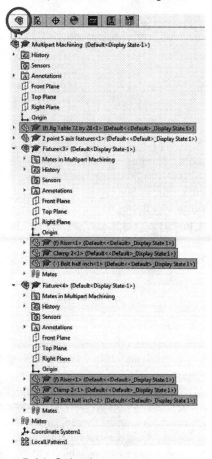

Figure 5.14 Selecting components to avoid

If you see a warning message: *The origin or machining direction or advanced parameters has changed, toolpaths need to be recalculated. Regenerate toolpaths now?* Click *Yes*. The toolpath will be regenerated like that shown in Figure 5.15.

5.4 The Sequence of Part Machining

You may click the *Simulate Toolpath* button to run the material removal simulation like that of Figure 5.1.

Note that the machining sequence follows a counterclockwise order looking down from above the jig table, starting from the part labeled 1 close to the front left corner, to part 10 close to the rear left corner, as shown in Figure 5.16.

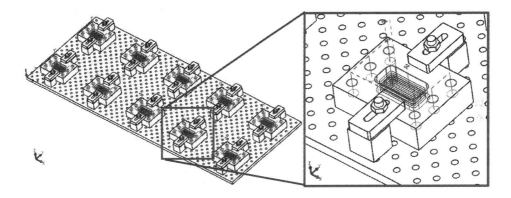

Figure 5.15 Toolpath of the multipart machining example

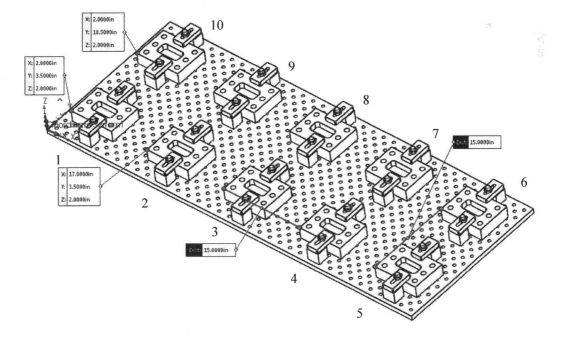

Figure 5.16 Locations of referencing corner points and distances between parts

The Rough Mill1 operation ran over all ten instances first, followed by Rough Mill2, Contour Mill, Center Drill, and Drill operations.

5.5   The G-code

Recall that in this example, we chose *Setup origin* for *Output origin* and *Fixture coordinate system* for *Setup origin* (see Figure 5.12). We expect that the G-code is output referring to the SOLIDWORKS assembly coordinate system, which is *Coordinate System1*, located at the front left corner of the top face of the jig table (as shown in Figure 5.10).

You may click the *Post Process* button ⬚ Post Process above the graphics area, and follow the same steps learned in the previous lessons (see, for example, Lesson 4) to convert the toolpaths into G-code.

Before reviewing the G-code, we acquire a basic understanding of the arrangement of the ten parts on the jig table. We also acquire a few referencing dimensions that help us verify the G-code.

As shown in Figure 5.16, the coordinates of the front left corner at the top face of stock 1 is (2, 3.5, 2), stock 2 is (17, 3.5, 2), and stock 10 is (2, 18.5, 2) referring back to *Coordinate System1*. The coordinates of the referencing points can be found by using the measure capability of SOLIDWORKS (by choosing from the pull-down menu *Tools > Evaluate > Measure*). Therefore, the distance between neighboring parts is 15in. along both X- and Y-directions, as shown in Figure 5.16.

Also, since the size of the block is 8in.×6in.×2in., the center points at the top face of the individual stocks are stock 1 (6, 6.5, 2), stock 2 (21, 6.5, 2), stock 3 (36, 6.5, 2), stock 4 (51, 6.5, 2), stock 5 (66, 6.5, 2), stock 6 (66, 21.5, 2), stock 7 (51, 21.5, 2), stock 8 (36, 21.5, 2), stock 9 (21, 21.5, 2), and stock 10 (6, 21.5, 2). All refer to *Coordinate System1*.

The X and Y coordinates of the hole centers, for example the three holes near the rear end of the pocket of stock 1, are respectively (3, 8.5), (3, 6.5), and (3, 4.5), from right to left [or Holes 1 to 3 shown in Figure 5.17(a)]. The other three holes near the front end of the pocket, from left to right, are respectively (9, 4.5), (9, 6.5), and (9, 8.5), or Holes 4 to 6 shown in Figure 5.17(b). All refer to *Coordinate System1*.

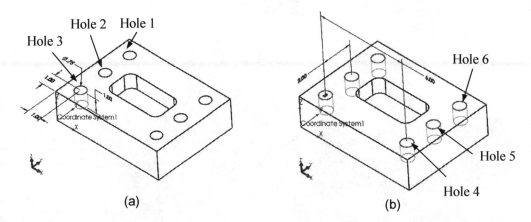

(a)   (b)

Figure 5.17 Dimensions of the hole centers of the design model: (a) the three holes near the rear end of the pocket, and (c) three holes near the front end of the pocket

```
O0001
N1 G20
N2 G91 G28 X0 Y0 Z0

N3 ( Rough Mill1 )
N4 (3/4 EM CRB 2FL 1-1/2 LOC)
N5 T04 M06
N6 S3677 M03
N7 G90 G54 G00 X4.985 Y6.515
N8 G43 Z2.1 H04 M08
N9 G01 Z1.625 F4.0448
N10 G17 Y6.485 F16.1793
N11 X7.015
N12 Y6.515
N13 X4.985
...

N107 ( Rough Mill1 )
N108 X19.985 Y6.515
N109 Z2.1
N110 G01 Z1.625 F4.0448
N111 Y6.485 F16.1793
N112 X22.015
N113 Y6.515
N114 X19.985
...

N208 ( Rough Mill1 )
N209 X34.985 Y6.515
N210 Z2.1
N211 G01 Z1.625 F4.0448
N212 Y6.485 F16.1793
N213 X37.015
N214 Y6.515
N215 X34.985
N216 Y6.815
N217 X34.685
N218 Y6.185
N219 X37.315
...

N1017 ( Rough Mill2 )
N1018 (1/2 EM CRB 2FL 1 LOC)
N1019 T03 M06
N1020 S6195 M03
N1021 G90 G54 G00 X4.5 Y7.24
N1022 G43 Z2.1 H03 M08
N1023 G01 Z1.75 F6.8151
N1024 G03 X4.26 Y7. I0 J-.24 F27.2606
N1025 G01 Y6.98
N1026 G02 X4.52 Y7.24 I.26 J0
N1027 G01 X4.5
N1028 G00 Z2.1
...

N1135 ( Rough Mill2 )
N1136 X19.5
N1137 Z2.1
N1138 G01 Z1.75 F6.8151
N1139 G03 X19.26 Y7. I0 J-.24 F27.2606
...

N2171 ( Contour Mill1 )
N2172 (3/4 EM CRB 2FL 1-1/2 LOC)
N2173 T04 M06
N2174 S3677 M03
N2175 G90 G54 G00 X6.3663 Y6.8273
N2176 G43 Z2.1 H04 M08
N2177 G01 Z1.625 F4.0448
N2178 G41 D24 X6.0905 Y7.103 F12.1345
```

```
N2179 G03 X6.0375 Y7.125 I-.053 J-.053
N2180 G01 X4.5 F16.1793
N2181 G03 X4.375 Y7. I0 J-.125
...

N2705 ( Center Drill1 )
N2706 (3/4 X 90DEG CBT SPOT DRILL)
N2707 T19 M06
N2708 S4991 M03
N2709 G90 G54 G00 X3. Y8.5
N2710 G43 Z3. H19 M08
N2711 G98 G82 Z1.6443 P00 R2.75 F26.4528
N2712 Y6.5
N2713 Y4.5
N2714 X9.
N2715 Y6.5
N2716 Y8.5
N2717 G80 Z3.

N2718 ( Center Drill1 )
N2719 X18.
N2720 G82 G98 R2.75 Z1.6443 P00 F26.4528
N2721 Y6.5
N2722 Y4.5
N2723 X24.
N2724 Y6.5
N2725 Y8.5
N2726 G80 Z3.

N2727 ( Center Drill1 )
N2728 X33.
N2729 G82 G98 R2.75 Z1.6443 P00 F26.4528
N2730 Y6.5
N2731 Y4.5
N2732 X39.
N2733 Y6.5
N2734 Y8.5
N2735 G80 Z3.

N2736 ( Center Drill1 )
N2737 X48.
N2738 G82 G98 R2.75 Z1.6443 P00 F26.4528
N2739 Y6.5
N2740 Y4.5
N2741 X54.
N2742 Y6.5
N2743 Y8.5
N2744 G80 Z3.
...

2781 ( Center Drill1 )
N2782 X18.
N2783 G82 G98 R2.75 Z1.6443 P00 F26.4528
N2784 Y21.5
N2785 Y19.5
N2786 X24.
N2787 Y21.5
N2788 Y23.5
N2789 G80 Z3.

N2790 ( Center Drill1 )
N2791 X3.
N2792 G98 G82 Z1.6443 P00 R2.75 F26.4528
N2793 Y21.5
N2794 Y19.5
N2795 X9.
N2796 Y21.5
N2797 Y23.5
N2798 G80 Z3. M09
N2799 G91 G28 Z0
```

```
N2800 ( Drill1 )
N2801 (25/32 SCREW MACH DRILL)
N2802 T20 M06
N2803 S4967 M03
N2804 G90 G54 G00 X3. Y8.5
N2805 G43 Z3. H20 M08
N2806 G98 G83 Z1. Q.1 R2.75 F29.8067
N2807 Y6.5
N2808 Y4.5
N2809 X9.
N2810 Y6.5
N2811 Y8.5
N2812 G80 Z3.

N2813 ( Drill1 )
N2814 X18.
N2815 G83 G98 R2.75 Z1. Q.1 F29.8067
N2816 Y6.5
N2817 Y4.5
N2818 X24.
N2819 Y6.5
N2820 Y8.5
N2821 G80 Z3.

N2822 ( Drill1 )
N2823 X33.
N2824 G83 G98 R2.75 Z1. Q.1 F29.8067
N2825 Y6.5
N2826 Y4.5
N2827 X39.
N2828 Y6.5
N2829 Y8.5
N2830 G80 Z3.

N2831 ( Drill1 )
N2832 X48.
N2833 G83 G98 R2.75 Z1. Q.1 F29.8067
N2834 Y6.5
N2835 Y4.5
N2836 X54.
N2837 Y6.5
N2838 Y8.5
N2839 G80 Z3.
...

N2876 ( Drill1 )
N2877 X18.
N2878 G83 G98 R2.75 Z1. Q.1 F29.8067
N2879 Y21.5
N2880 Y19.5
N2881 X24.
N2882 Y21.5
N2883 Y23.5
N2884 G80 Z3.

N2885 ( Drill1 )
N2886 X3.
N2887 G83 G98 R2.75 Z1. Q.1 F29.8067
N2888 Y21.5
N2889 Y19.5
N2890 X9.
N2891 Y21.5
N2892 Y23.5
N2893 G80 Z3. M09
N2894 G91 G28 Z0
N2895 G28 X0 Y0
N2896 M30
```

Figure 5.18 The G-code, with no subroutine

Figure 5.18 shows (partial) contents of the G-code (O0001), consisting of over 2800 blocks, which is large. This is because every single operation is repeated ten times for the respective ten instances.

One easier way to verify the G-code is to review the cutter locations of the center drill and drill operations. The NC blocks of center drill and drill operations are listed in the second and third columns of Figure 5.18, respectively.

NC blocks N2705 to N2717 center drill the six holes of stock 1. Block N2709 moves the drill to Hole 1 (G00 X3. Y8.5), N2712 center drills Hole 2 (X3. Y6.5), and N2713 to N2716 center drill Holes 3 to 6, respectively.

NC blocks N2718 to N2726 center drill the six holes of stock 2. Block N2719 moves the drill to Hole 1 (X18. Y8.5), N2721 center drills Hole 2 (X18. Y6.5), and N2722 to N2725 center drill Holes 3 to 6, respectively.

You may review more NC blocks to find the cutter locations for the center drill operations on the remaining 8 stocks.

The NC blocks of drill operations are listed in the third column of Figure 5.18, respectively. The cutter locations of the drill operations are identical to those of center drills as they should be.

The G-code output by the CAMWorks post processor M3AXIS-TUTORIAL seems to be all good.

Although the G-code shown in Figure 5.18 is good, the code is lengthy. It is desirable to output G-code as subroutines for machining operations that cut part instances.

In CAMWorks, the option to output G-code with subroutines can be found under the *Posting* tab of the *Machine* dialog box (see Figure 5.19).

You may right click *Machine* in the CAMWorks operation tree 📐 and choose *Edit Definition* to bring up the *Machine* dialog box. Choose the *Posting* tab. Click the checkbox in front of *Output subroutines for part instances and feature patterns* under *Subroutines* (circled in Figure 5.19) and click *OK*.

Click the *Post Process* button 📄 to convert the toolpath into G-code again. The G-code now consists of one main program: O0001, and five subroutines. They are O0002 (Rough Mill1), O0003 (Rough Mill2), O0004 (Contour Mill1), O0005 (Center Drill), and O0006 (Drill).

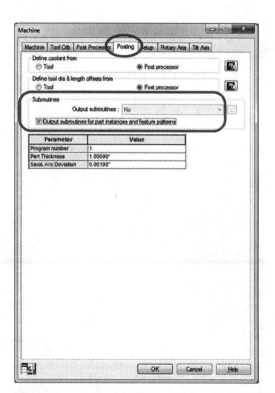

Figure 5.19 The *Posting* tab of the *Machine* dialog box

As shown in Figure 5.20, in addition to a few blocks at the beginning and the end, the main program is written roughly in five segments, Rough Mill1 (N3 to N47), Rough Mill2 (N48 to N92), Contour Mill1 (N93 to N137), Center Drill1 (N138 to N182), and Drill1 (N183 to N229).

In the first segment, Rough Mill1, a local coordinate system was assigned at the center point of the top face of individual stocks using the G52 NC word; for example, blocks N8 (stock 1, X6. Y6.5 Z2.), N12 (stock 2, X21. Y6.5 Z2.), to N44 (stock 10, X6. Y21.5 Z2.). Subprogram O0002 was called using M98 NC word right after the local coordinate system was set; for example, blocks N9 (stock 1), N13 (stock 2), to N45 (stock 10). G52 appears again in the main program—for example, N10, N14, N18, to N46, right after returning from subprogram to set the coordinate system back to (0,0,0).

The same can be found for the remaining operations: Rough Mill2, Contour Mill1, Center Drill1, and Drill1.

Next, we take a look at the subroutines, in particular Center Drill1 (O0005) and Drill1 (O0006), shown in the last column of Figure 5.21.

In Center Drill1 operation (O0005), the locations of the center of the six holes are specified in blocks N1 (Hole1: X-3. Y2.) and N4-N8 referring to the center points of Holes 2-6 at top face of the respective stock. Note that since the part setup origin was set at the top left corner of the jig table, the main program sets local coordinate systems to the center point of the top face of the respective stock by referring back to *Coordinate System1*. This is how the post process was written.

```
O0001                          N48 ( Rough Mill2 )            N138 ( Center Drill1 )
N1 G20                         N49 (1/2 EM CRB 2FL 1 LOC)     N139 (3/4 X 90DEG CBT SPOT
N2 G91 G28 X0 Y0 Z0            N50 T03 M06                    DRILL)
                               N51 S6195 M03                  N140 T19 M06
N3 ( Rough Mill1 )             N52 G90                        N141 S4991 M03
N4 (3/4 EM CRB 2FL 1-1/2 LOC)  N53 G52 X6. Y6.5 Z2.           N142 G90
N5 T04 M06                     N54 M98 P0003                  N143 G52 X6. Y6.5 Z2.
N6 S3677 M03                   N55 G52 X0 Y0 Z0               N144 M98 P0005
N7 G90                         ...                            N145 G52 X0 Y0 Z0
N8 G52 X6. Y6.5 Z2.                                           ...
N9 M98 P0002
N10 G52 X0 Y0 Z0               N88 ( Rough Mill2 )
                               N89 G52 X6. Y21.5 Z2.          N178 ( Center Drill1 )
N11 ( Rough Mill1 )            N90 M98 P0003                  N179 G52 X6. Y21.5 Z2.
N12 G52 X21. Y6.5 Z2.          N91 G52 X0 Y0 Z0               N180 M98 P0005
N13 M98 P0002                  N92 G91 G28 Z0                 N181 G52 X0 Y0 Z0
N14 G52 X0 Y0 Z0                                              N182 G91 G28 Z0
                               N93 ( Contour Mill1 )
N15 ( Rough Mill1 )            N94 (3/4 EM CRB 2FL 1-1/2 LOC) N183 ( Drill1 )
N16 G52 X36. Y6.5 Z2.          N95 T04 M06                    N184 (25/32 SCREW MACH DRILL)
N17 M98 P0002                  N96 S3677 M03                  N185 T20 M06
N18 G52 X0 Y0 Z0               N97 G90                        N186 S4967 M03
...                            N98 G52 X6. Y6.5 Z2.           N187 G90
                               N99 M98 P0004                  N188 G52 X6. Y6.5 Z2.
N43 ( Rough Mill1 )            N100 G52 X0 Y0 Z0              N189 M98 P0006
44 G52 X6. Y21.5 Z2.           ...                           N190 G52 X0 Y0 Z0
N45 M98 P0002                                                 ...
N46 G52 X0 Y0 Z0              N133 ( Contour Mill1 )
N47 G00 G91 G28 Z0           N134 G52 X6. Y21.5 Z2.          N223 ( Drill1 )
                             N135 M98 P0004                  N224 G52 X6. Y21.5 Z2.
                             N136 G52 X0 Y0 Z0               N225 M98 P0006
                             N137 G91 G28 Z0                 N226 G52 X0 Y0 Z0
                                                             N227 G91 G28 Z0
                                                             N228 G28 X0 Y0
                                                             N229 M30
```

Figure 5.20 The G-code, main program

Similarly, for Drill1 operations, the main program (third column in Figure 5.20) sets local coordinate systems (G52) and calls subprogram O0006 (M98). The contents of subprogram O0006 shown in the last column of Figure 5.21 are similar to those of subprogram O0005.

For the pocket milling operations, as shown in the first two columns of Figure 5.20, the main program sets local coordinate systems (G52) and calls subprogram O0002 for Rough Mill1, O0003 for Rough Mill2, and then O0004 for Contour Mill1 (M98). The partial contents of subprograms O0002, O0003, and O0004 are shown in the first three columns of Figure 5.21. The X- and Y-coordinates of the cutter locations shown in the subprograms O0002, O0003, and O0004 are referred to the respective local coordinate systems, which are again the center point at the top face of the respective stocks, set by using G52 in the main program.

We have verified that G-code is correctly generated. We have now completed this exercise. You may save your model for future references.

O0002	O0003	O0004	O0005
N1 G90 G54 G00 X-1.015 Y.015	N1 G90 G54 G00 X-1.5 Y.74	N1 G90 G54 G00 X.3663 Y.3273	N1 G90 G54 G00 X-3. Y2.
N2 G43 Z1. H04 M08	N2 G43 Z1. H03 M08	N2 G43 Z1. H04 M08	N2 G43 Z1. H17 M08
N3 Z.1	N3 Z.1	N3 Z.1	N3 G82 G98 R.75 Z-.3557 P00
N4 G01 Z-.375 F4.0448	N4 G01 Z-.25 F6.8151	N4 G01 Z-.375 F4.0448	F26.4528
N5 G17 Y-.015 F16.1793	N5 G17 G03 X-1.74 Y.5 I0 J-.24	N5 G41 D24 X.0905 Y.603 F12.1345	N4 Y0
N6 X1.015	F27.2606	N6 G17 G03 X.0375 Y.625 I-.053 J-.053	N5 Y-2.
N7 Y.015	N6 G01 Y.48	N7 G01 X-1.5 F16.1793	N6 X3.
N8 X-1.015	N7 G02 X-1.48 Y.74 I.26 J0	N8 G03 X-1.625 Y.5 I0 J-.125	N7 Y0
N9 Y.315	N8 G01 X-1.5	N9 G01 Y-.5	N8 Y2.
N10 X-1.315	N9 G00 Z.1	N10 G03 X-1.5 Y-.625 I.125 J0	N9 G80 Z1. M09
N11 Y-.315	N10 Z-.15	N11 G01 X1.5	N10 M99
N12 X1.315	N11 G01 Z-.5 F6.8151	N12 G03 X1.625 Y-.5 I0 J.125	
N13 Y.315	N12 G03 X-1.74 Y.5 I0 J-.24 F27.2606	N13 G01 Y.5	
N14 X-1.015	...	N14 G03 X1.5 Y.625 I-.125 J0	O0006
N15 Y.615		...	N1 G90 G54 G00 X-3. Y2.
N16 X-1.5	N108 Z-.65		N2 G43 Z1. H18 M08
...	N109 G01 Z-1. F6.8151	N47 G01 Y.5	N3 G83 G98 R.75 Z-1. Q.1 F29.8067
	N110 G02 X1.74 Y.48 I0 J-.26 F27.2606	N48 G03 X1.5 Y.625 I-.125 J0	N4 Y0
N96 G03 X1.615 Y-.5 I0 J.115	N111 G01 Y.5	N49 G01 X-.0375	N5 Y-2.
N97 G01 Y.5	N112 G03 X1.5 Y.74 I-.24 J0	N50 G03 X-.0905 Y.603 I0 J-.075	N6 X3.
N98 G03 X1.5 Y.615 I-.115 J0	N113 G01 X1.48	N51 G40 G01 X-.3663 Y.3273	N7 Y0
N99 G01 X-1.015	N114 G00 Z.1	N52 G00 Z.1	N8 Y2.
N100 G00 Z.1	N115 Z1. M09	N53 Z1. M09	N9 G80 Z1. M09
N101 Z1. M09	N116 M99	N54 M99	N10 M99
N102 M99			

Figure 5.21 The G-code of subprograms, O0002, O0003, O0004, O0005, and O0006

5.6 Exercises

Problem 5.1. Create an assembly like that of Figure 5.22 using the parts (*Jig table*, *Problem 5.1 Part*, *Clamp*, and *Short Bolt*) and subassembly (*Fixture*) in the Exercises folder of Lesson 5 downloaded from the publisher's website. Note that *Problem 5.1 Part* is identical to that of Problem 4.1. Create a total of eight instances as a linear pattern feature.

(a) Generate a machining simulation for the eight instances using the machining operations created in Problem 4.1.

(b) Generate G-code with selections like those discussed in this lesson. Verify that the codes are generated correctly by reviewing the contents of the codes, similar to those of Section 5.5.

Hint: You may perform the following to manually create a machinable feature in assembly, (a) create instances by right clicking *Part Manager* and choosing *Manage Parts*, (b) insert a mill part setup by expanding *Problem 5.1 Part* and right clicking *Feature Manager*, and (c) insert a new multi surface feature by expanding *Feature Manager* and right clicking *Mill Part Setup*.

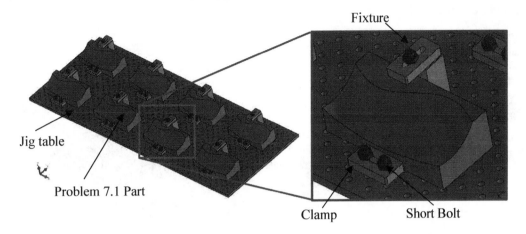

Figure 5.22 The assembly model of Problem 5.1

Problem 5.2. Now we get back to the example of this lesson to explore other options in selecting part setup origin. Bring out the *Setup Parameters* dialog box and select *Part setup origin* for *Output origin* (see Figure 5.23). Note that the part setup origin symbols appear at the front left corner at the top face of individual stocks, as shown in Figure 5.24.

Output G-code. Open the G-code file and identify the differences that this selection makes to the G-code ouput in Section 5.5. Verify if the G-code are output correctly.

Figure 5.23 The *Origin* tab in the *Setup Parameters* dialog box

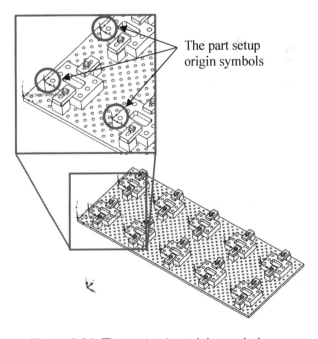

Figure 5.24 The part setup origin symbols

Problem 5.3. Continue from Problem 5.2. In this case, we choose *Setup origin* for *Output origin* (see Figure 5.25), click *Entity select* under *Setup origin*, and pick *Coordinate System1* shown in Figure 5.26. Then, we output G-code. Open the G-code and identify the differences that such options make to the G-code output in Section 5.5. Verify if the G-code are output correctly.

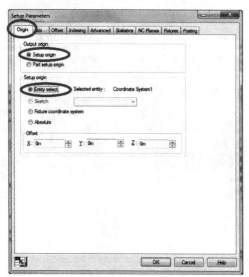

Figure 5.25 The *Origin* tab in the
Setup Parameters dialog box

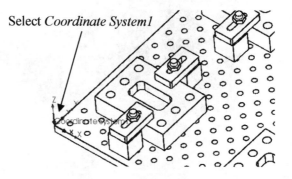

Select *Coordinate System1*

Figure 5.26 Select a corner point at part
setup origin

Lesson 6: Multiplane Machining

6.1 Overview of the Lesson

In this lesson, we learn to create machining operations that cut parts with machinable features on multiple planes. In particular, we assume the stocks to be mounted on a tombstone that rotates with desired angles by a rotary table rotating along the longitudinal direction; that is, the 4th axis or A-axis in this case.

The design part is similar to that of Lessons 3 and 5, except that there are holes on the side faces, in addition to the pocket and six holes on the top face. Similar to Lesson 5, the face milling discussed in Lesson 3 is not included to simplify the fixture design for stock setup. The part is assembled to a tombstone by using two identical fixtures (one on each end). Each fixture consists of a clamp and two bolts. A total of four parts mounted on the respective four faces of the tombstone are to be machined, as shown in Figure 6.1. The tombstone is mounted on a rotary table using four bolts, and the rotary table is assembled to a rotary unit, which is mounted on a jig table like that of Lesson 5 (not shown in Figure 6.1). We assume that a 3-axis mill with a rotary table is employed for this lesson. The rotary table rotates ±90° or ±180° along the longitudinal direction for the tools to cut features on faces that are perpendicular to the tool axis.

We will learn the steps to define the 4th axis, in addition to creating instances of part for generating machining operations that are similar to that of Lesson 5, including defining stock for individual instances, extracting machinable features, generating toolpath, and selecting components in the assembly for the tools to avoid. We will take a closer look at the G-code generated by CAMWorks to verify that the code is generated correctly in support of the machining operations.

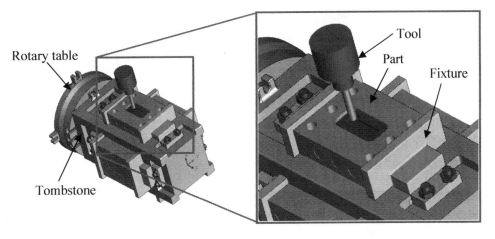

Figure 6.1 Material removal simulation of the multiplane machining example

6.2   The Multiplane Machining Example

A design part (filename: *2 point 5 axis features with side holes.SLDPRT*) similar to that of Lessons 3 and 5 with a bounding box of 8in.×6in.×2in. is employed in this lesson. In addition to the pocket and holes on its top face—see Figure 6.2(a)—there are three holes on its front side face, as shown in Figure 6.2(a), and another three holes on its rear side face; see Figure 6.2(b). All these holes on the side faces are of the same size, and they are extracted as machinable features by using the automatic feature recognition (AFR) capability.

The assembly model (*Multiplane Machining.SLDASM*) consists of eight parts, one subassembly (called *clamped part*), a circular pattern feature that includes the instances of the clamped part subassembly, and a point and a coordinate system, as listed in the FeatureManager design tree 🍀 shown in Figure 6.3(a).

The jig table of Lesson 5 is employed again for this lesson. A rotary unit is mounted on the jig table; see Figure 6.3(b). And a rotary table is assembled to the rotary unit by using four 0.5in. bolts, and a tombstone on which parts are mounted is assembled to the rotary table. The clamped part subassembly consisting of the design part (*2 and 5 axis features with side holes*), two fixtures, and two bolts are mounted on the respective four faces of the tombstone; see Figure 6.3(c).

A coordinate system, *Coordinate System1*, is defined at the center of the front end face of the tombstone with X-axis pointing along the longitudinal direction of the jig table; see Figure 6.3(b). Note that *Coordinate System1* will be chosen as the fixture coordinate system for the machining operations in this lesson. The rotary table provides rotation motion along the X-axis of the fixture coordinate system.

You may open the example file with toolpaths created (filename: *Multiplane Machining with Toolpath.SLDASM*) to preview the setups and toolpaths of this example.

You may click the CAMWorks feature tree tab 🔲 to preview the setups. There are four setups corresponding to the respective four tool axes defined in this example. As shown in Figure 6.4(a), the tool axis (symbol: ⊗) of *Setup1* points in the –Z direction with four machinable features: *Rectangular Pocket1* and *Hole Group1* on the top face of the part mounted on the top face of the tombstone, *Hole Group2* on the side face of the part mounted on the front side face of the tombstone, and *Hole Group3* on the part mounted on the rear side face.

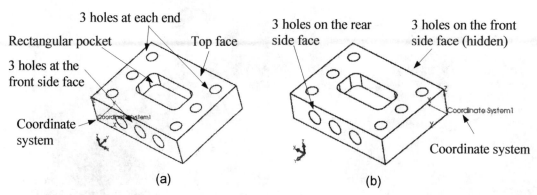

Figure 6.2 Machinable features of the design part, *2 and 5 axis features with side holes*, (a) features at the top and front side faces, and (b) holes on the rear side face (a rotated view)

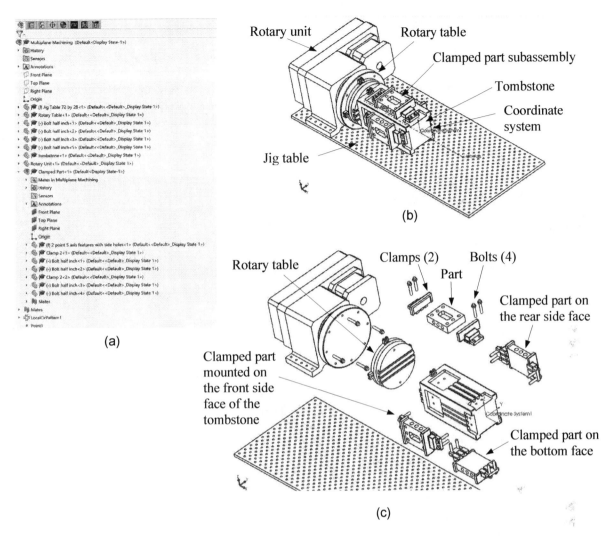

Figure 6.3 The solid model of multiplane machining assembly example: (a) model tree, (b) entire assembly, and (c) explode view

As another example, the tool axis of *Setup3* points in the +Y direction—see Figure 6.4(b)—with another four machinable features, *Rectangular Pocket1* and *Hole Group1* on the part mounted on the front side face of the tombstone, *Hole Group2* on the part mounted on the bottom face, and *Hole Group3* on the part of the top face of the tombstone.

Tool axes of *Setup2* and *Setup4* point in +Z and –Y directions, respectively. Machinable features associated with these two setups are pockets and holes sketched on the part faces that are normal to the respective tool axes, similar to those of *Setup1* and *Setup3* shown in Figure 6.4.

Note that you might see the four setups in a different order as in the example file when you go through the exercise of this lesson.

The stock size is identical to that of the bounding box of the part. The pocket milling operations, (including two rough milling and a contour milling) and hole drilling operations (center drill and drill) are identical to those of Lesson 5 (for example, the first five operations of *Setup1* shown in Figure 6.5).

Similarly, the drilling operations for holes of parts mounted on the side faces of the tombstone consist of a center drill and a drill operation (see Figure 6.5). The toolpaths of *Setup1* are shown in Figure 6.5 for illustration. Similar toolpaths can be found for the other three setups. Note that the part setup origin coincides with that of the origin of *Coordinate System1*, as pointed out in Figure 6.5. You may click the CAMWorks operation tree tab ![icon], expand the *Part Manager* node, expand the part (*2 point 5 axis features with side holes.SLDPRT*), and then expand the *Instances* to see the four instances of the part to be cut (*Clamped part-1* to *Clamped part-4* shown in Figure 6.6).

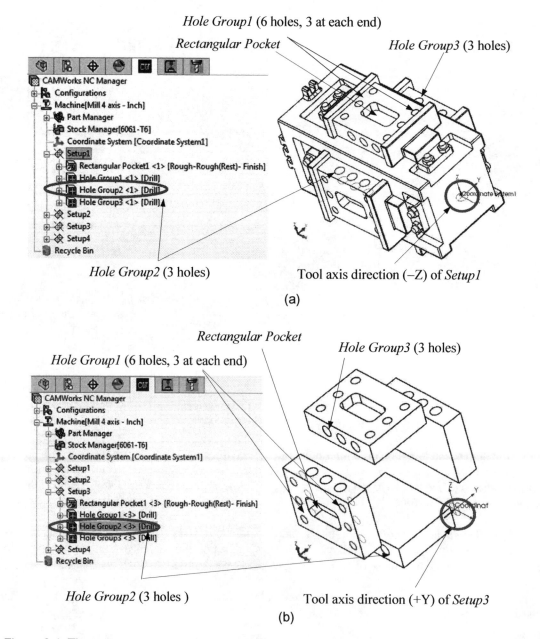

Hole Group1 (6 holes, 3 at each end)

Rectangular Pocket

Hole Group3 (3 holes)

Hole Group2 (3 holes)

Tool axis direction (–Z) of *Setup1*

(a)

Rectangular Pocket

Hole Group1 (6 holes, 3 at each end)

Hole Group3 (3 holes)

Hole Group2 (3 holes)

Tool axis direction (+Y) of *Setup3*

(b)

Figure 6.4 The setups generated for the multiplane machining example: (a) *Setup1* and associated machinable features, and (b) *Setup3* and associated machinable features (tombstone and fixtures not shown)

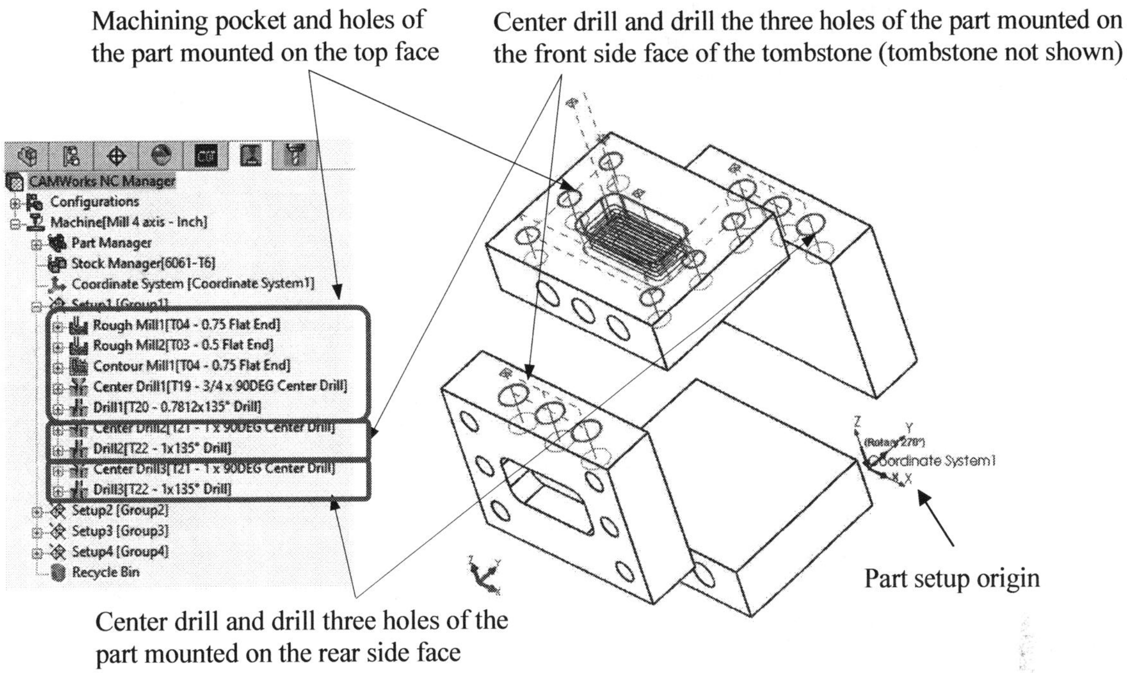

Machining pocket and holes of the part mounted on the top face

Center drill and drill the three holes of the part mounted on the front side face of the tombstone (tombstone not shown)

Part setup origin

Center drill and drill three holes of the part mounted on the rear side face

Figure 6.5 Toolpaths of the nine operations of *Setup1* (tombstone and fixtures not shown)

You may click the *Simulate Toolpath* button ⏚ above the graphics area to preview the machining operations.

6.3   Using CAMWorks

Open SOLIDWORKS Assembly

Open the assembly model (filename: *Multiplane Machining.SLDASM*) downloaded from the publisher's website. This assembly model appears in the graphics area similar to that of Figure 6.3(b). Again, as soon as you open the model, you may want to check the unit system chosen and make sure the IPS system is selected. You may also increase the decimals from the default 2 to 4 digits similar to that of previous lessons.

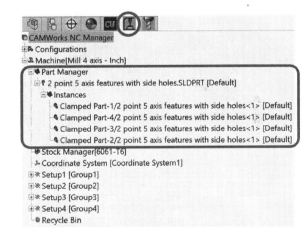

Figure 6.6 The instances and NC operations listed under the CAMWorks operation tree tab

Select NC Machine

Click the CAMWorks feature tree tab ▣ and right click *Mill-inch* to select *Edit Definition*. In the *Machine* dialog box, we choose *Mill 4 axis-inch* under the *Machine* tab; see Figure 6.7(a). Click *Select*.

Choose *Tool Crib 2* under *Available tool cribs* of the *Tool Crib* tab, select *M4AXIS-TUTORIAL* under the *Post Processor* tab—Figure 6.7(b), and select *Coordinate System1* under the *Fixture Coordinate System* of the *Setup* tab.

Define the Rotary Axis

In the *Machine* dialog box, click the *Rotary Axis* tab, choose *X axis* as the *Rotary axis* and *XZ plane* as the *0 degree position*—circled in Figure 6.8(a). Then, click *OK*. In the graphics area, a circular arc with a counterclockwise arrow appears at the origin of *Coordinate System1*, as shown in Figure 6.8(b), indicating the rotation direction of the rotary axis. Click *OK* to accept the machine definition.

Manage Part

Similar to Lesson 5, we are dealing with an assembly of multiple design parts. Therefore, we have to assign which part or parts to cut. Under the CAMWorks feature tree tab ▨, right click *Part Manager* and select *Manage Parts* (Figure 6.9). The *Manage Parts* dialog box appears (Figure 6.10).

Pick the design part (*2 point 5 axis features with side holes*) in the graphics area or expand *Clamped Part*, and select *2 point 5 axis features with side holes* under the FeatureManager design tree tab ▨. The part is now listed under *Selected Parts* in the *Manage Parts* dialog box (Figure 6.10). Click the part, and click *Add All Instances* button to bring in all instances. Click *OK* to accept the part definition.

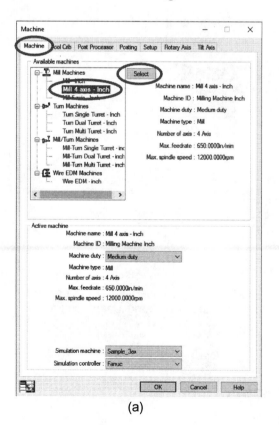
(a)

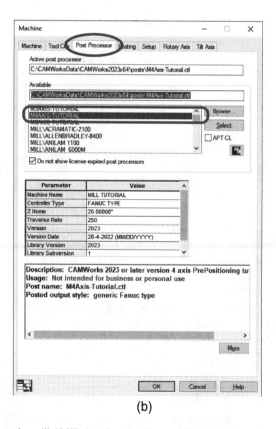
(b)

Figure 6.7 The *Machine* dialog box, (a) selecting 4 axis mill (*Mill 4 axis - inch*), and (b) selecting the 4 axis post-processor (*M4AXIS-TUTORIAL*)

Click the CAMWorks feature tree tab ![CW], expand the *Part Manager* node to see the part (*2 point 5 axis features with side holes.SLDPRT*) and its instances to make sure that all four parts are included, as shown in Figure 6.11. Also, a *Stock Manager* node has been added.

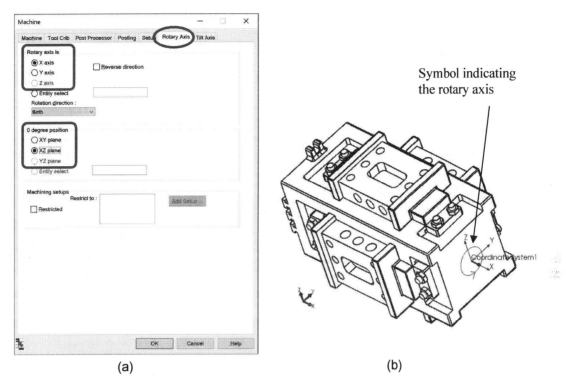

Symbol indicating
the rotary axis

(a)   (b)

Figure 6.8 Defining the rotary axis, (a) the *Rotary Axis* tab of the *Machine* dialog box, and (b) the
symbol indicating the rotary axis

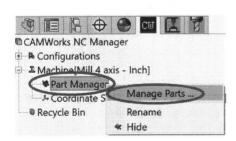

Figure 6.9 Right clicking *Part
Manager* and select *Manage Parts*

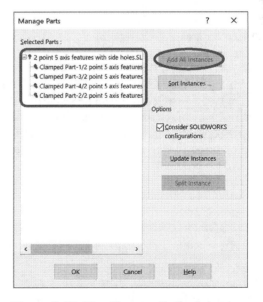

Figure 6.10 The *Manage Parts* dialog box

Create Stock for Part Instances

From CAMWorks feature tree tab 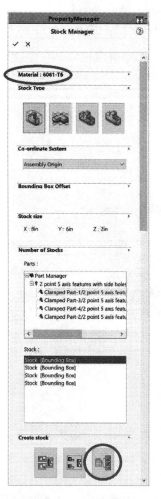, right click *Stock Manager* and choose *Edit Definition*. In the *Stock Manager* dialog box (Figure 6.12), we leave the default stock size (8in.×6in.×2in.) and choose *6061-T6* for stock material, and click the *Apply Current Definitions to All Parts* button (circled in Figure 6.12). Then click the checkmark ✔ to accept the definition of stock.

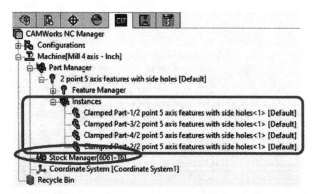

Figure 6.11 All four instances of the target part listed under CAMWorks feature tree

Figure 6.12 The *Stock Manager* dialog box

Extract Machinable Features

Click the *Extract Machinable Features* button above the graphics area. Four setup entities (*Setup1-4*) are created with four machinable features extracted per setup: *Rectangular Pocket1*, *Hole Group1* (with six holes), *Hole Group2* (with three holes), and *Hole Group3* (with another three holes), listed in CAMWorks feature tree (see Figure 6.13). All machinable features are shown in magenta color.

Click a setup to locate its part setup origin and tool axis in the graphics area. For example, click *Setup2* to see a symbol with arrow pointing upward (coinciding with the +Z-axis) appearing at the center of the front end face of the tombstone (coinciding with *Coordinate System1*, as shown in Figure 6.14). All four setups shared the same part setup origin but with different tool axes. Tool axes of the four setups are pointing respectively in –Z, +Z, +Y, and –Y directions. Again, you may see setups with a different order. The order does not affect the end results, as long as you have all four setups, including ±Z and ±Y, in place.

Generate Operation Plan and Toolpath

Click the *Generate Operation Plan* button above the graphics area. Nine operations, *Rough Mill1*, *Rough Mill2*, *Contour Mill1*, *Center Drill1*, *Drill1*, *Center Drill2*, *Drill2*, *Center Drill3*, and *Drill3*, are generated for each setup. There is a total of 36 operations, nine per setup. They are listed in CAMWorks operation tree (see Figure 6.15, showing operations of *Setup1* and *Setup4*). Again they are shown in magenta color. Click the *Generate Toolpath* button 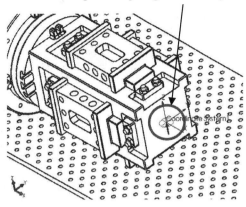 above the graphics area to create the toolpath. The operations are turned into black color after toolpaths are generated.

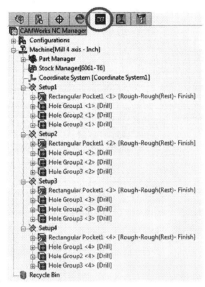

Figure 6.13 The machinable features extracted

The part setup origin chosen by default

Figure 6.14 The default part setup origin coinciding with *Coordinate System1*

Fixtures and Components to Avoid

Similar to Lesson 5, we select components to avoid in generating the toolpath. In this lesson, we select the rotary table, tombstone, fixtures, and bolts to avoid. We will not select the rotary unit and the jig table since the chance that the tools collide with them is minimum for a 3-axis mill with a rotary table.

On the other hand, only the components selected as fixtures are included in the material removal simulation. Excluding the jig table and the rotary unit makes the simulation more visually logical and realistic.

Click the CAMWorks operation tree tab , and right click *Setup1* to choose *Edit Definition*. In the *Setup Parameters* dialog box, click the *Features* tab (Figure 6.16). Choose the components to be part of the fixtures, including the rotary table, tombstone, both clamps, and bolts, either from the graphics area or from the FeatureManager design tree tab (see Figure 6.17) by clicking them.

Figure 6.15 The nine operations generated per setup

Also, click both clamps and bolts from the remaining three sets, including those of the circular pattern feature of the solid model. You may expand the pattern feature (*LocalCirPattern1* under FeatureManager design tree) to select the clamps and bolts. After selecting all these components, we click the *Avoid All* button to avoid them in the toolpath generation, then we click *OK*.

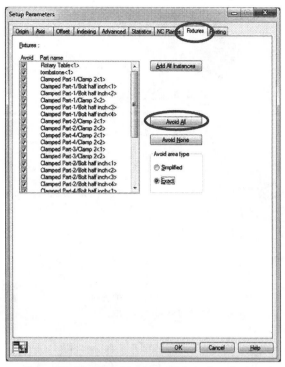

Figure 6.16 The *Features* tab in the *Setup Parameters* dialog box

Figure 6.17 Selecting the components to avoid

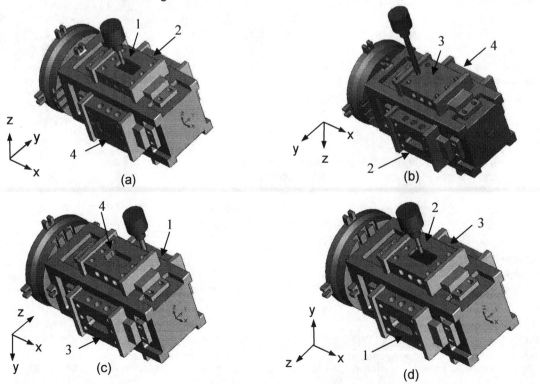

Figure 6.18 Material removal simulation (stock rotates), (a) *Setup1*, (b) *Setup2*, (c) *Setup3*, and (d) *Setup4*

Click *Yes* to the question in the warning box: *The origin or machining direction or advanced parameters has changed, toolpaths need to be recalculated. Regenerate toolpaths now?* The toolpaths will be regenerated, for example, like that shown in Figure 6.5 for *Setup1*.

Note that sometimes undesirable toolpaths may be generated by CAMWorks, for example, cutters reaching areas outside the pocket, as occasionally observed in Rough Mill2 and Contour Mill1 of *Setup1*. All you have to do is to regenerate toolpath by right clicking these NC operations and selecting *Generate Toolpath*.

6.4   The Sequence of Part Machining

You may click the *Simulate Toolpath* button ![Simulate Toolpath icon] to run the material removal simulation like that of Figure 6.1.

Note that the machining sequence follows that of the four setups, from *Setup1* to *Setup4*. The rotary table rotates along the X-axis of *Coordinate System1*. The nine machining operations of *Setup1* cut machinable features on parts labeled 1, 2, and 4, as shown in Figure 6.18(a).

The first three operations cut the pocket (2 rough and 1 contour mills) on part 1, followed by two operations (center drill and hole drilling) that drill the six holes on the top face of part 1. Then two operations drill the three holes (center drill and hole drilling) on part 4. The final two operations cut the three holes on part 2. The nine operations repeat three more times for the remaining respective three setups, as shown in Figure 6.18(b), (c), and (d) for *Setup2*, *Setup3*, and *Setup4*, respectively.

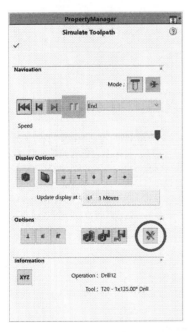

Figure 6.19 Click the *Options* button in the *Toolpath Simulation* toolbox

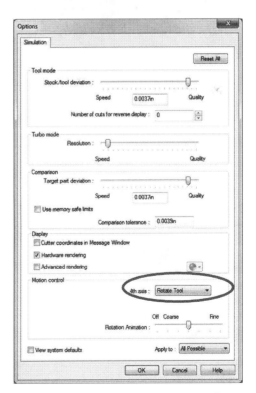

Figure 6.20 The *Features* tab in the *Setup Parameters* dialog box

In the default setting of the material removal simulation, stock rotates in the material removal simulation. The rotary table rotates a –180° angle from *Setup1* to *Setup2*, a 90° angle from *Setup2* to *Setup3*, and then a 180° angle from *Setup3* to *Setup4*.

You may change the default setting to rotate the tool in the material removal simulation (although this is less desirable from a practical perspective).

This can be done by clicking the *Options* button ▧ in the *Toolpath Simulation* toolbox (Figure 6.19). In the *Options* dialog box (Figure 6.20), choose *Rotate Tool* for *4th axis* under *Motion control*, and then click *OK*.

Click *Yes* to the message: *Simulation must be restarted before new settings will take effect. Do you want to restart simulation?*

Click the *Run* button in the *Toolpath Simulation* toolbox. The material removal simulation will take place for all four setups, in which the tools rotate like those of Figure 6.21.

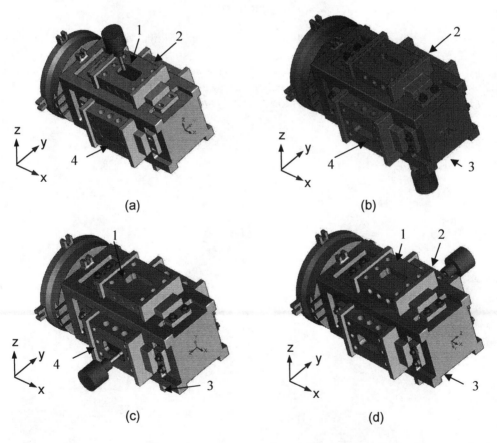

Figure 6.21 Material removal simulation (tool rotates), (a) *Setup1*, (b) *Setup2*, (c) *Setup3*, and (d) *Setup4*

6.5   Reviewing The G-code

In this example, we chose the part setup origin at the center point of the front end face of the tombstone, coinciding with that of the fixture coordinate system, *Coordinate System1*, as shown in Figure 6.4. We expect that the G-code generated by CAMWorks refers to the part setup origin at the coordinate system. In addition, recall that we did not choose *Output subroutines for part instances and feature pattern* as we did in Lesson 5 while outputting the G-code (see Figure 5.19 of Lesson 5). We expect that the G-code generated does not include subprogram calls. Therefore, the main program includes NC blocks that perform the nine machining operations four times for the respective four setups following the order shown in Figure 6.18.

To understand the G-code, we first locate the center points of a few selected holes. We use the *Measure* option (*Tools > Evaluate > Measure*) of SOLIDWORKS and choose *Center to Center* option; see Figure 6.22(a).

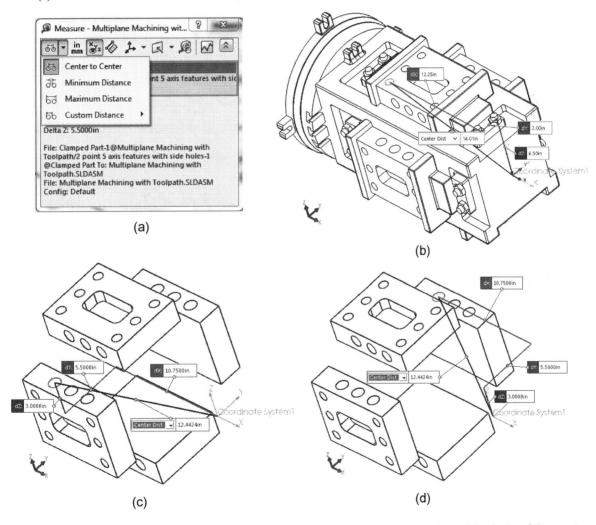

Figure 6.22 Coordinates of selected hole centers, (a) the *Measure* dialog box, (b) a hole of the part mounted on the top face of the tombstone selected, (c) a selected hole of the part mounted on the front side face of the tombstone (tombstone not shown), and (d) another hole of the part on the rear side face (tombstone not shown)

We first select the coordinate system (*Coordinate System1*) and pick the boundary edge of one of the holes of the part mounted on top; see Figure 6.22(b) for the hole selected. The coordinates of the center of the selected hole can be located at (−12.25, −2, 6.5), as shown in Figure 6.22(b). Since the distance between the neighboring holes is 2in., the coordinates of the center points of the remaining two holes close to the rear end of the top face are, respectively, (−12.25, 0, 6.5), and (−12.25, 2, 6.5).

Similarly, we select the coordinate system (*Coordinate System1*) and pick the boundary edge of one of the holes of the part mounted on the front side face; see Figure 6.22(c) for the hole picked. The coordinates of the hole center are (−10.75, −5.5, 3.0), as shown in Figure 6.22(c). Since the distance between the neighboring holes is 1.75in., the coordinates of the center points of the remaining two holes are, respectively, (−9.0, −5.5, 3.0), and (−7.25, −5.5, 3.0).

Also, we select the coordinate system (*Coordinate System1*) and pick the boundary edge of the hole of the part mounted on the rear side of the tombstone; see Figure 6.22(d). The coordinates of the hole center are (−10.75, 5.5, 3.0), as shown in Figure 6.22(d). Since the distance between the neighboring holes is 1.75in., the center points of the remaining holes are, respectively, (−9.00, 5.5, 3.0), and (−7.25, 5.5, 3.0).

You may click the *Post Process* button ⬚ above the graphics area, and follow the same steps learned in Lesson 4 to convert the toolpath into G-code. Figure 6.23 shows (partial) contents of the NC program (O0001).

As shown in Figure 6.23, the NC program is organized in four segments, led by identifier A defining the rotation angle of the 4th axis for the respective four setups. They are NC blocks *N7 A−90*, *N360 A−270*, *N713 A−180*, and *N1066 A0*, representing *Setup1* to *Setup4*, respectively. These rotation angles are consistent with those shown in Figure 6.18 except that a −90° angle is in place before *Setup1*. This is because the XZ plane was selected as zero degree position, as shown in Figure 6.8(a).

Each segment consists of nine operations. These operations, for example for *Setup1* (N7: A = −90), start at blocks *N3 (Rough Mill1)*, *N108 (Rough Mill2)*, *N227 (Contour Mill1)*, *N284 (Center Drill1)*, *N298 (Drill1)*, *N312 (Center Drill2)*, *N323 (Drill2)*, *N334 (Center Drill3)*, and *N345 (Drill3)*, for the respective operations.

We take a closer look at the *Center Drill1* operation. Blocks N288 and N289 move the center drill bit 7.5in. above the center point of the first hole (X = −12.25, Y = 2.0) on the top face of the part—the third hole to the right of the one selected in Figure 6.22(b). Block N290 makes a center drill to the hole, then to the next hole in the middle in N291 (X= −12.25, Y = 0), and then to the third hole in N292 (X = −12.25, Y = −2.0). This operation can also be seen by using the *Step Through Toolpath* option ⬚, for example, blocks N284 to N297 for *Center Drill1*, as shown in Figure 6.24(a).

Also shown are blocks N312 to N322 for *Center Drill2* [Figure 6.24(b)], blocks N334 to N344 for *Center Drill3* [Figure 6.24(c)], and blocks N1040 to N1050 shown in Figure 6.24(d) for *Center Drill9* of *Setup3* with 4th axis: A*−180*.

The complete content of the G-code can be seen in the file: *Multiplane Machining no Subroutines.txt*.

We have now completed the lesson. You may save your model for future references.

```
O0001
N1 G20
N2 G91 G28 X0 Y0 Z0

N3 ( Rough Mill1 )
N4 (3/4 EM CRB 2FL 1-1/2 LOC)
N5 T04 M06
N6 S3677 M03
N7 G90 G54 G00 X-10.265 Y.015 A-90. B0
N8 G43 Z6.6 H04 M08
N9 G01 Z6.125 F4.0448
N10 Y-.015 F16.1793
N11 X-8.235
N12 Y.015
N13 X-10.265
N14 Y.315
N15 X-10.565
N16 Y-.315
N17 X-7.935
N18 Y.315
...

N108 ( Rough Mill2 )
N109 (1/2 EM CRB 2FL 1 LOC)
N110 T03 M06
N111 S6195 M03
N112 G90 G54 G00 X-10.75 Y.74
N113 G43 Z6.6 H03 M08
N114 G01 Z6.25 F6.8151
N115 G03 X-10.99 Y.5 I0 J-.24 F27.2606
N116 G01 Y.48
N117 G02 X-10.73 Y.74 I.26 J0
...

N227 ( Contour Mill1 )
N228 (3/4 EM CRB 2FL 1-1/2 LOC)
N229 T04 M06
N230 S3677 M03
N231 G90 G54 G00 X-8.8837 Y.3273
N232 G43 Z6.6 H04 M08
N233 G01 Z6.125 F4.0448
N234 G41 D24 X-9.1595 Y.603 F12.1345
N235 G03 X-9.2125 Y.625 I-.053 J-.053
N236 G01 X-10.75 F16.1793
...

N284 ( Center Drill1 )
N285 (3/4 X 90DEG CBT SPOT DRILL)
N286 T19 M06
N287 S4991 M03
N288 G90 G54 G00 X-12.25 Y2.
N289 G43 Z7.5 H19 M08
N290 G98 G82 Z6.1443 P00 R6.6 F26.4528
N291 Y0
N292 Y-2.
N293 X-6.25
N294 Y0
N295 Y2.
N296 G80 Z7.5 M09
N297 G91 G28 Z0

N298 ( Drill1 )
N299 (25/32 SCREW MACH DRILL)
N300 T20 M06
N301 S4967 M03
N302 G90 G54 G00 X-12.25 Y2.
N303 G43 Z7.5 H20 M08
N304 G98 G83 Z5.5 Q.1 R6.6 F29.8067
N305 Y0
...

N312 ( Center Drill2 )
N313 (1 X 90DEG CBT SPOT DRILL)
N314 T21 M06
N315 S3880 M03
N316 G90 G54 G00 X-10.75 Y-5.5
N317 G43 Z4. H21 M08
N318 G98 G82 Z2.55 P00 R3.1 F23.285
N319 X-9.
N320 X-7.25
N321 G80 Z4. M09
N322 G91 G28 Z0
```

```
N323 ( Drill2 )
N324 (1 SCREW MACH DRILL)
N325 T22 M06
N326 S3880 M03
N327 G90 G54 G00 X-10.75 Y-5.5
N328 G43 Z4. H22 M08
N329 G98 G83 Z1.75 Q.1 R3.1 F23.285
N330 X-9.
...

N334 ( Center Drill3 )
N335 (1 X 90DEG CBT SPOT DRILL)
N336 T21 M06
N337 S3880 M03
N338 G90 G54 G00 X-10.75 Y5.5
N339 G43 Z4. H21 M08
N340 G98 G82 Z2.55 P00 R3.1 F23.285
N341 X-9.
N342 X-7.25
N343 G80 Z4. M09
N344 G91 G28 Z0

N345 ( Drill3 )
N346 (1 SCREW MACH DRILL)
N347 T22 M06
N348 S3880 M03
N349 G90 G54 G00 X-10.75 Y5.5
N350 G43 Z4. H22 M08
N351 G98 G83 Z1.75 Q.1 R3.1 F23.285
...

N356 ( Rough Mill3 )
N357 (3/4 EM CRB 2FL 1-1/2 LOC)
N358 T04 M06
N359 S3677 M03
N360 G90 G54 G00 X-10.265 Y.015 A-270.
N361 G43 Z6.6 H04 M08
N362 G01 Z6.125 F4.0448
...

N461 ( Rough Mill4 )
N462 (1/2 EM CRB 2FL 1 LOC)
N463 T03 M06
N464 S6195 M03
N465 G90 G54 G00 X-10.75 Y.74
N466 G43 Z6.6 H03 M08
N467 G01 Z6.25 F6.8151
N468 G03 X-10.99 Y.5 I0 J-.24 F27.2606
N469 G01 Y.48
N470 G02 X-10.73 Y.74 I.26 J0
N471 G01 X-10.75
...

N580 ( Contour Mill2 )
N581 (3/4 EM CRB 2FL 1-1/2 LOC)
N582 T04 M06
N583 S3677 M03
N584 G90 G54 G00 X-8.8837 Y.3273
N585 G43 Z6.6 H04 M08
...

N665 ( Center Drill5 )
N666 (1 X 90DEG CBT SPOT DRILL)
N667 T21 M06
N668 S3880 M03
N669 G90 G54 G00 X-10.75 Y-5.5
N670 G43 Z4. H21 M08
N671 G98 G82 Z2.55 P00 R3.1 F23.285
N672 X-9.
...

N698 ( Drill6 )
N699 (1 SCREW MACH DRILL)
N700 T22 M06
N701 S3880 M03
N702 G90 G54 G00 X-10.75 Y5.5
N703 G43 Z4. H22 M08
N704 G98 G83 Z1.75 Q.1 R3.1 F23.285
N705 X-9.
N706 X-7.25
N707 G80 Z10. M09
N708 G91 G28 Z0
```

```
N709 ( Rough Mill5 )
N710 (3/4 EM CRB 2FL 1-1/2 LOC)
N711 T04 M06
N712 S3677 M03
N713 G90 G54 G00 X-10.265 Y.015 A-180.
N714 G43 Z6.6 H04 M08
N715 G01 Z6.125 F4.0448
N716 Y-.015 F16.1793
N717 X-8.235
...

N1040 ( Center Drill9 )
N1041 (1 X 90DEG CBT SPOT DRILL)
N1042 T21 M06
N1043 S3880 M03
N1044 G90 G54 G00 X-10.75 Y5.5
N1045 G43 Z4. H21 M08
N1046 G98 G82 Z2.55 P00 R3.1 F23.285
N1047 X-9.
N1048 X-7.25
N1049 G80 Z4. M09
N1050 G91 G28 Z0

N1051 ( Drill9 )
N1052 (1 SCREW MACH DRILL)
N1053 T22 M06
N1054 S3880 M03
N1055 G90 G54 G00 X-10.75 Y5.5
N1056 G43 Z4. H22 M08
N1057 G98 G83 Z1.75 Q.1 R3.1 F23.285
N1058 X-9.
N1059 X-7.25
N1060 G80 Z10. M09
N1061 G91 G28 Z0

N1062 ( Rough Mill7 )
N1063 (3/4 EM CRB 2FL 1-1/2 LOC)
N1064 T04 M06
N1065 S3677 M03
N1066 G90 G54 G00 X-10.265 Y.015 A0
N1067 G43 Z6.6 H04 M08
N1068 G01 Z6.125 F4.0448
N1069 Y-.015 F16.1793
N1070 X-8.235
N1071 Y.015
...

N1167 ( Rough Mill8 )
N1168 (1/2 EM CRB 2FL 1 LOC)
N1169 T03 M06
N1170 S6195 M03
N1171 G90 G54 G00 X-10.75 Y.74
N1172 G43 Z6.6 H03 M08
N1173 G01 Z6.25 F6.8151
N1174 G03 X-10.99 Y.5 I0 J-.24 F27.2606
...

N1393 ( Center Drill12 )
N1394 (1 X 90DEG CBT SPOT DRILL)
N1395 T21 M06
N1396 S3880 M03
N1397 G90 G54 G00 X-10.75 Y5.5
N1398 G43 Z4. H21 M08
N1399 G98 G82 Z2.55 P00 R3.1 F23.285
N1400 X-9.
N1401 X-7.25
N1402 G80 Z4. M09
N1403 G91 G28 Z0

N1404 ( Drill12 )
N1405 (1 SCREW MACH DRILL)
N1406 T22 M06
N1407 S3880 M03
N1408 G90 G54 G00 X-10.75 Y5.5
N1409 G43 Z4. H22 M08
N1410 G98 G83 Z1.75 Q.1 R3.1 F23.285
N1411 X-9.
N1412 X-7.25
N1413 G80 Z4. M09
N1414 G91 G28 Z0
N1415 G28 X0 Y0
N1416 M30
```

Figure 6.23 The G-code generated by CAMWorks (partial contents)

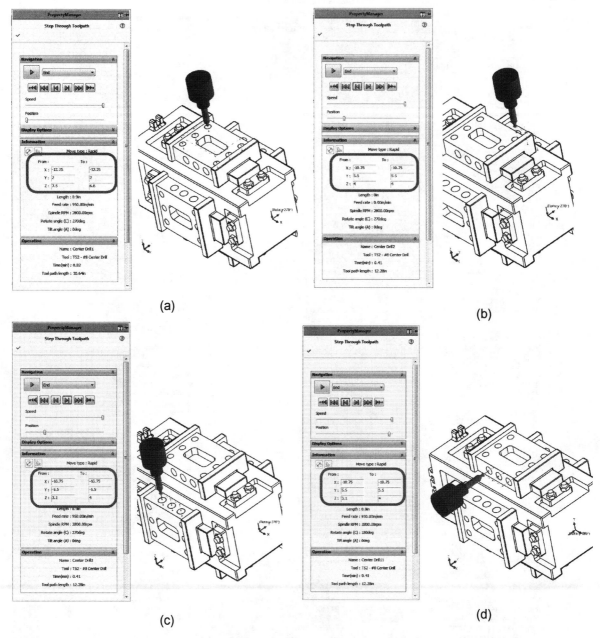

Figure 6.24 Step through the toolpath: (a) N284 to N297 (Center Drill1), (b) starting blocks N312 (Center Drill2), (c) starting blocks N334 (Center Drill3), and (d) starting blocks N1040 (Center Drill9, *Setup3*)

6.6   Exercises

Problem 6.1. Create an assembly like that of Figure 6.25 using the parts (*Rotary Table*, *Problem 6.1 Part*, *Clamp 2*, and *Bolt half inch*) and subassembly (*Clamped Part*) in the Problem 6.1 folder downloaded from the publisher's website. Note that the freeform surface of *Problem 6.1 Part* is similar to that of Lesson 4. In addition to the freeform surface, there are holes at the top and side faces of the part. Create a total of four instances as a circular pattern feature.

(a) Generate machining operations for the four instances following steps discussed in this lesson. The operations must cut the freeform surface and holes on the top and side faces. Note that like Lesson 4, you will have to manually create a multi-surface machinable feature to machine the freeform surface.

(b) Generate G-code and verify that the codes are generated correctly by reviewing the contents of the codes, similar to those of Section 6.5.

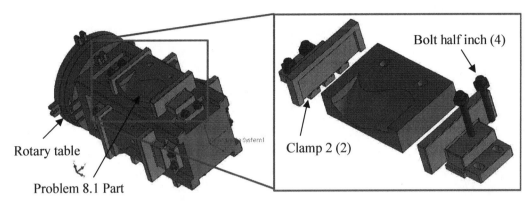

Figure 6.25 Assembly of Problem 6.1

Problem 6.2. Go back to the example in this lesson and select the *Output subroutines for part instances and feature patterns* option in the *Machine* dialog box (see Figure 6.26, and also see the discussion on Figure 5.19 of Lesson 5). Then, output G-code. Does the selection output G-code with subprogram calls in machining the instances? Verify if the G-code output from CAMWorks are correct.

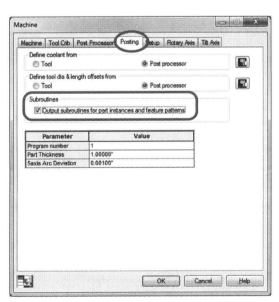

Figure 6.26 The *Posting* tab of the
Machine dialog box

[Notes]

Lesson 7: Multiaxis Milling and Machine Simulation

7.1 Overview of the Lesson

In this lesson, we learn to create machining operations that cut a design part with a freeform surface using a 5-axis mill. The freeform surface of the design part is, in this lesson, a cylindrical surface extruded with a Bézier curve of four control points. In addition to creating a 5-axis milling operation, we will learn to avoid excessive cuts by turning on gouge checking. Moreover, we will learn to bring the toolpath simulation into a virtual CNC machine to carry out the virtual machining simulation in a more realistic setting. As mentioned in Lesson 1, CAMWorks offers Machine Simulation capability that employs virtual CNC machines to support users in carrying out machining simulation. In this lesson, we will use *Mill_Tutorial*, one of the generic virtual machines that come with CAMWorks, to learn Machine Simulation capability. In addition to simulating machining operations, Machine Simulation supports tool collision detection in a more realistic setting. More importantly, this capability allows you to add your own virtual CNC machine, a virtual replica of the physical machine available in your machine shop, into CAMWorks to carry out virtual machining simulation. We will discuss the steps of adding an external mill, a 5-axis HAAS mill, to CAMWorks in Lesson 10.

We create three operations to cut the part, similar to those of Lesson 4, including a volume milling, a local milling, and a surface milling operation. In Lesson 4 we assumed a 3-axis mill. In this lesson, we use 5-axis mill for surface milling, as shown in Figure 7.1(a). Thereafter, we change the surface milling sequence using a larger flat-end cutter to discuss the capability in CAMWorks that adjusts the toolpath to avoid gouging. The toolpath will also be simulated using a legacy machine, *Mill_Tutorial*, in Machine Simulation, as shown in Figure 7.1(b). This legacy machine is similar to the work chamber of a typical 5-axis mill with a tilt table and a rotary table that provide rotations in X and Z-axes, respectively; for example, a HAAS VF-5 CNC mill shown in Figure 7.1(c).

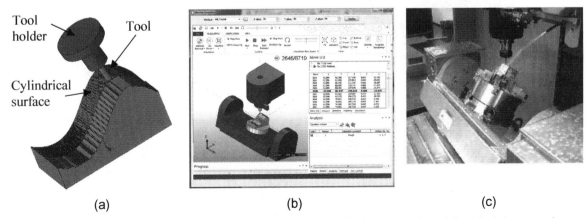

(a) (b) (c)

Figure 7.1 Toolpath simulation of the multiaxis machining operation, (a) material removal simulation, (b) CAMWorks Machine Simulation, and (c) HAAS VF-5 CNC mill

At the end of this lesson, we re-visit the surface milling operation of the freeform surface example in Lesson 4 and explore the feasibility of using a 5-axis surface milling operation to improve the accuracy and quality of the machined surface.

7.2   The Multiaxis Machining Example

The design part (filename: *Cylindrical Surface.SLDPRT*) consists of a solid extrusion feature—see Figure 7.2(a)—with a sketch of three straight lines and a Bézier curve of four control points, as shown in Figure 7.2(b). These four control points labeled P_0, P_1, P_2, and P_3 are dimensioned so that their X-Y coordinates are defined on the sketch plane as (0, 3), (4, 9), (5, –2) and (10, 2), respectively. The size of the bounding box of the solid feature is 10in.×4.833in.×5in. The stock employed for this lesson is a rectangular block made of Steel 1005, of 10in.×5.083in.×5in.; that is, a 0.25in. extension in the +Y direction from its bounding box. The extrusion of the Bézier curve generates a cylindrical surface, in which the quality of the machined surface would be desirable by implementing the three-operation scenario similar to that of Lesson 4 but with a 5-axis mill for the surface milling operation.

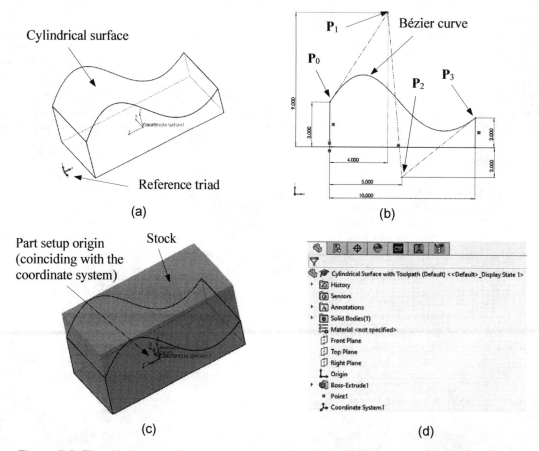

Figure 7.2 The design part, *Cylindrical Surface.SLDPRT*, (a) the solid model and the cylindrical surface, (b) the Bézier curve in sketch, (c) stock and part setup origin, and (d) the FeatureManager design tree

A coordinate system, *Coordinate System1*, is defined with its origin at the center of the bottom face of the solid model (*Point1*). Like previous lessons, *Coordinate System1* is chosen as the fixture coordinate system, again defining the "home point" or main zero position on the machine. Unlike previous lessons, the X-, Y-, and Z-axes of *Coordinate System1* point in different directions as those of the reference triad shown at the lower left corner of the graphics area. X-, Y-, and Z-axes of the fixture coordinate system must be defined consistently with those of the virtual machine in Machine Simulation, in which X- and Z-axes align with those of the tilt table and rotary table, respectively. This will become clear in Section 7.7 when we discuss Machine Simulation. The origin of the coordinate system will be chosen as the part setup origin; see Figure 7.2(c). Again, a part setup origin defines the G-code program zero location.

The unit system chosen is IPS (inch, pound, second). When you open the solid model *Cylindrical Surface.SLDPRT*, you should see the solid feature (*Boss-Extrude1*), a point, and a coordinate system listed in the FeatureManager design tree 🎨 like that of Figure 7.2(d).

There are three operations to be created for this example. The first operation is a volume milling (*Area Clearance*), which is a rough cut using a 2in. flat-end mill. The toolpath of the volume milling operation can be seen in Figure 7.3(a). The second operation is a local milling (another area clearance operation) that continues removing material remaining from the volume milling using a smaller ball-nose cutter of diameter 0.5in.; see toolpath in Figure 7.3(b). The third operation is a multiaxis surface milling serving as a finish cut that intends to improve the quality and accuracy of the machined part to meet the requirements; see toolpath in Figure 7.3(c).

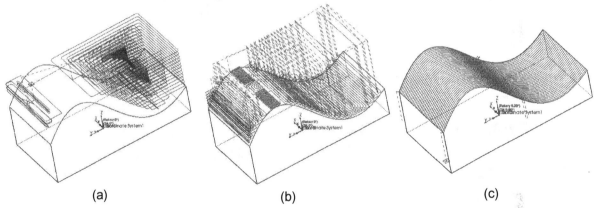

(a)   (b)   (c)

Figure 7.3 Toolpath of the three NC operations: (a) volume milling, (b) local milling, and (c) multiaxis surface milling

You may open the example file with toolpath created (filename: *Cylindrical Surface with toolpath.SLDPRT*) to preview the toolpaths of this example. When you open the file, you should see the four operations listed under CAMWorks operation tree tab 📕 , as shown in Figure 7.4. Note that the operation *Multiaxis Surface Milling Gouging* is suppressed. You may un-suppress it to review a toolpath that avoids gouging. You may simulate individual operations by right clicking the entity and choosing *Simulate Toolpath*, or simulate the combined operations by clicking the *Simulate Toolpath* button ⬡ Simulate Toolpath above the graphics area.

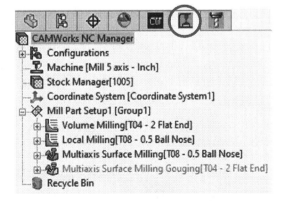

Figure 7.4 The NC operations listed under the CAMWorks operation tree tab

You may also right click *Mill Part Setup1* and choose *Machine Simulation > Legacy* to bring up the *Machine Simulation* window like that of Figure 7.1(b).

7.3   Using CAMWorks

Open SOLIDWORKS Part

Open the part file (filename: *Cylindrical Surface.SLDPRT*) downloaded from the publisher's website. This solid model, as shown in Figure 7.2(d), consists of one extrude solid feature, a point, and a coordinate system. As soon as you open the model, you may want to check the unit system chosen and make sure the IPS system is selected. You may also increase the decimals from the default 2 to 4 digits similar to that of the previous lesson. Since steps in creating the first two operations are similar to those of Lesson 4, some screen captures of dialog boxes will be skipped to minimize repetitions.

Select NC Machine

Click the CAMWorks feature tree tab and right click *Mill-inch* to select *Edit Definition*. In the *Machine* dialog box, we select *Mill 5 axis-inch* under *Machine* tab—Figure 7.5(a)—and click *Select*, choose *Tool Crib 2* under *Available tool cribs* of the *Tool Crib* tab, select *M5AXIS-TUTORIAL* under the *Post Processor* tab, Figure 7.5(b), and select *Coordinate System1* under *Fixture Coordinate System* of the *Setup* tab.

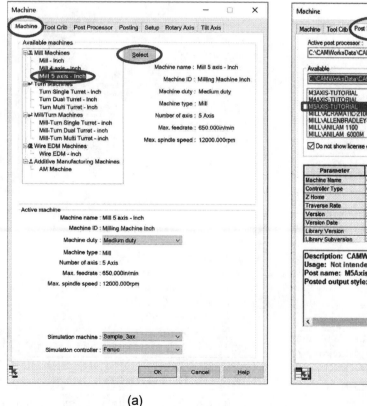

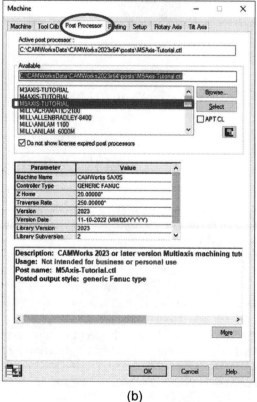

(a)   (b)

Figure 7.5 The *Machine* dialog box, (a) selecting the 5-axis mill, and (b) selecting the 5-axis post-processor

Next, we choose Z-axis as the rotary axis, and X-axis as the tilt axis.

In the *Machine* dialog box, we choose the *Rotary Axis* tab, and select *Z axis* as the rotary axis; see Figure 7.6(a). Then, we choose the *Tilt Axis* tab, and select *X axis* as the tilt axis, and *XY plane* for the *0 degree position*; see Figure 7.6(b). We click *OK* to accept the definition.

Create Stock

From CAMWorks feature tree , right click *Stock Manager* and choose *Edit Definition*. In the *Stock Manager* dialog box, we increase the Y+ field from 0 to 0.25in. and choose *1005* for stock material. A rectangular stock of 10in.×5.0827in.×5in. should appear in the graphics area similar to that of Figure 7.2(c).

Next, we select the cylindrical surface to create a machinable feature manually since the cylindrical surface is not a standard 2.5 axis features recognizable by AFR.

Create a Machinable Feature

Click the CAMWorks feature tree tab , right click *Stock Manager*, and choose *Mill Part Setup*, as shown in Figure 7.7.

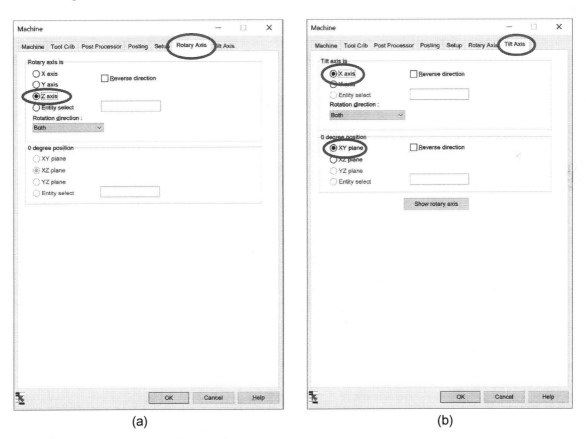

(a)   (b)

Figure 7.6 Defining axes of the 5-axis mill, (a) rotary axis: Z, and (b) tilt axis: X

The *Mill Setup* dialog box appears (Figure 7.8), select the *Top Plane* and click the *Reverse Selected Entity* button 🔁 under *Entity* to reverse the direction. Make sure that the arrow of the tool axis symbol ⬦ points in a downward direction, as shown in Figure 7.8. Click the checkmark ✔ to accept the definition. A *Mill Part Setup1* is now listed in the feature tree.

Now we define a machinable feature. From CAMWorks feature tree 🖼 , right click *Mill Part Setup1* and choose *Multi Surface Feature* (Figure 7.9). The *Multi Surface Feature* dialog box appears (Figure 7.10).

Pick the cylindrical surface of the part in the graphics area; the surface picked is now listed under *Selected Faces*. Leave the default *Area Clearance, Pattern Project* for *Strategy*, and click the checkmark ✔ to accept the machinable feature.

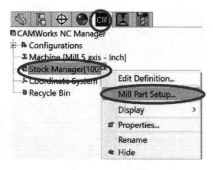

Figure 7.7 Selecting *New Mill Part Setup*

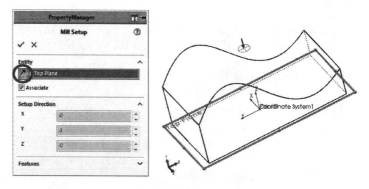

Figure 7.8 Picking the Top Plane for mill setup

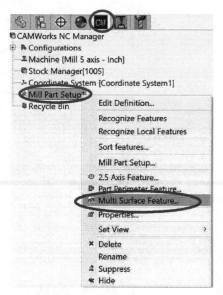

Figure 7.9 Choosing *New Multi Surface Feature*

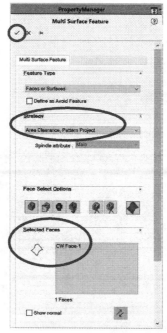

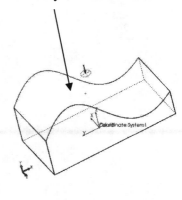

Pick the cylindrical surface

Figure 7.10 Picking the cylindrical surface for machinable feature

In the CAMWorks feature tree 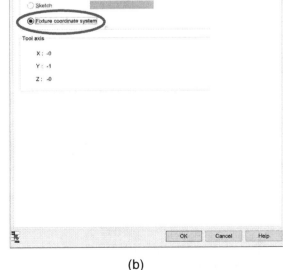, a *Multi Surface Feature1* is added in magenta color under *Mill Part Setup1*.

Generate Operation Plan and Toolpath

Click the *Generate Operation Plan* button above the graphics area. Two operations, *Area Clearance1* and *Pattern Project1*, are generated. They are listed in CAMWorks operation tree . Again, they are shown in magenta color. Change the name of the operation from *Area Clearance* to *Volume Milling*.

The part setup origin and axis coincide with the fixture coordinate system by default. We will stay with the default selection. You may review them by right clicking *Mill Part Setup1* and choosing *Edit Definition*.

In the *Part Setup Parameters* dialog box, see Figure 7.11(a); the *Fixture coordinate system* under the *Origin* tab has been chosen. Click the *Axis* tab, as shown in Figure 7.11(b); again, the *Fixture coordinate system* has been chosen. The default selections set the part setup origin and axes to coincide with the coordinate system, *Coordinate System1*, as shown in Figure 7.2(c). Click *OK* to close the dialog box.

Click the *Generate Toolpath* button above the graphics area to create the toolpath. The two operations are turned into black color right after toolpaths are generated.

We change the tool of the *Volume Milling* operation to a 2in. flat-end cutter. This can be done by right clicking *Volume Milling* and choosing *Edit Definition*.

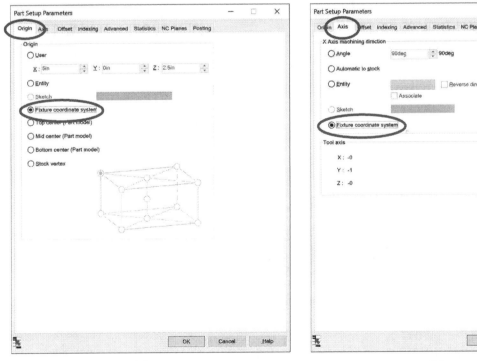

(a)   (b)

Figure 7.11 The *Part Setup Parameters* dialog box, (a) the *Origin* tab, and (b) the *Axis* tab

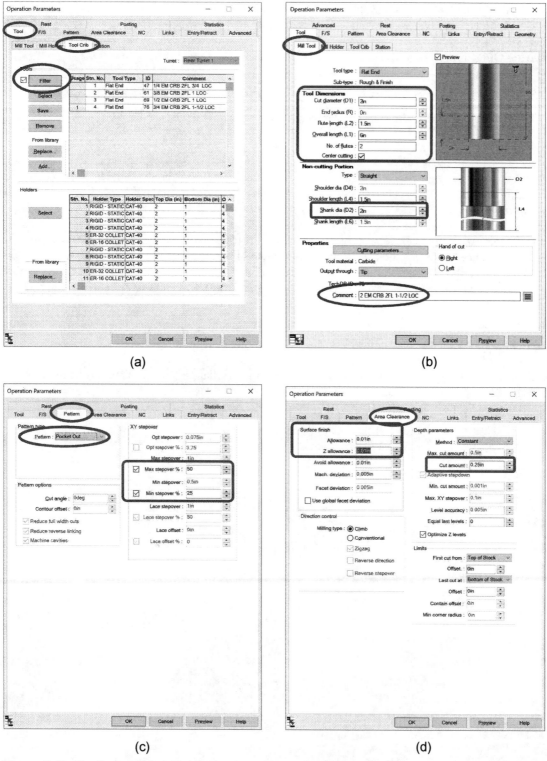

(a)   (b)

(c)   (d)

Figure 7.12 Replacing with a 2in. flat-end cutter, (a) showing flat-end cutters in the tool crib using the filter option, (b) 2in. flat-end cutter selected, (c) *Pattern type* and *XY stepover* under the *Pattern* tab, and (d) *Allowance* and *Cut amount* under the *Area Clearance* tab

In the *Operation Parameters* dialog box, as shown in Figure 7.12(a), choose *Tool Crib* tab and click *Filter* to display flat-end tools only. Note that there is no 2in. flat-end cutter in the tool crib. We will have to add a new tool to the tool library via CAMWorks TechDB™, and then add it to the tool crib. Before learning CAMWorks TechDB™ in Lesson 10, we will simply modify the size of the cutter, increasing from ¾ to 2in.

Choose the *Mill Tool* tab, and enter *2* for *Cut diameter (D1)*, *2* for *Shank dia (D2)*, *6* for overall length, and enter *2 EM CRB 2FL 1-1/2 LOC* for Comment. The modified cutter appears under the *Tool* tab; see Figure 7.12(b).

Choose the *Pattern* tab, choose *Pocket Out* for *Pattern type*, and enter *50%* and *25%* for *Max.* and *Min. stepover*, respectively, as shown in Figure 7.12(c).

Choose the *Area Clearance* tab and enter *0.01* for the *Allowance* and *Z allowance* under *Surface finish*, as shown in Figure 7.12(d). The allowances leave 0.01in. thickness material on the bottom face (and wall of pocket if cutting a pocket). This thin layer material minimizes (or eliminates) tool marks on the finished surface that improves the quality of the machined part. Enter *0.25in.* for *Cut amount*.

We leave the remaining machining parameter values determined by the TechDB™ for this exercise.

Click *OK* to accept the changes and click *Yes* to the warning message: *Operation parameters have changed, toolpaths need to be recalculated. Regenerate toolpaths now?*

The toolpath will be generated like that shown in Figure 7.3(a).

Right click *Volume Milling* and choose *Simulate Toolpath* to show the material removal simulation of the operation, like that of Figure 7.13(a). As shown in Figure 7.13(a), there is a significant amount of material uncut. Next, we add a local milling operation to further remove the material remaining on the cylindrical surface. Also, we delete *Pattern Project1* by right clicking it and choosing *Delete*. We will create a multiaxis surface milling operation shortly.

7.4   Adding a Local Milling Operation

Similar to Lesson 4, we add a local milling operation by creating another area clearance operation that continues machining the material remaining from the *Volume Milling* operation with a ball-nose cutter of 0.5in. diameter, and 0.2in. for both stepover and depth of cut.

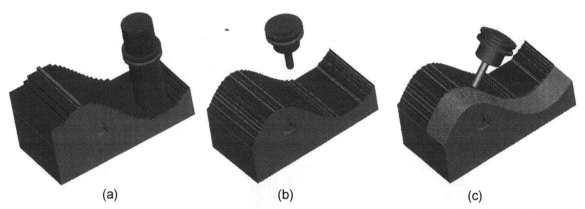

(a)   (b)   (c)

Figure 7.13 Material removal simulation of the three NC operations, (a) volume milling, (b) local milling, and (c) multiaxis surface milling

Right click *Mill Part Setup1* and choose *3 Axis Mill Operations > Area Clearance*.

The *New Operation: Area Clearance* dialog box (Figure 7.14) appears. Under the *Tool* tab, select *T08 – 0.5 Ball Nose*; a sketch of the tool with dimensions appears to the right (see Figure 7.14).

Click the *Features* tab, select *Multi Surface Feature1 [Area Clearance, Pattern Project]*, and click the checkmark ✔ to accept the new operation (see Figure 7.15).

A new operation, *Area Clearance2*, is now listed in the CAMWorks feature tree tab ▨ (see Figure 7.16) and the *Operation Parameters* dialog box appears.

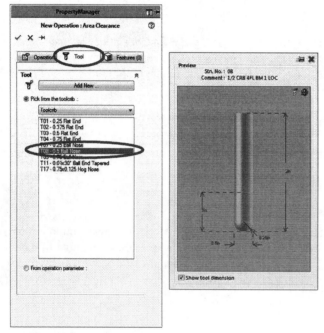

Figure 7.14 The *New Operation: Area Clearance* dialog box

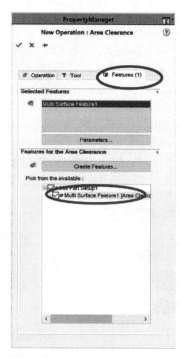

Figure 7.15 The *Features* tab of the *New Operation: Area Clearance* dialog box

Choose the *Pattern* tab of the *Operation Parameters* dialog box, select *Pocket Out* for *Pattern*, and enter *40%* and *25%* for *Max stepover%* and *Min stepover%*, as shown in Figure 7.17(a).

Choose the *Area Clearance* tab of the *Operation Parameters* dialog box. In the *Depth* parameters group, enter *0.01in* for both *Allowance* and *Z Allowance*, and enter *0.2in* for *Cut amount*, as shown in Figure 7.17(b).

Choose the *Rest* tab of the *Operation Parameters* dialog box, as shown in Figure 7.18(a). Choose *From WIP* for *Method*, and click the selection button ▢ to select *Volume Milling* for *Compute WIP from operations*; see Figure 7.18(b). This is how we link the local milling to a previous operation.

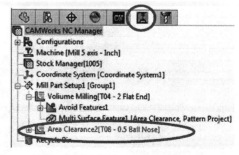

Figure 7.16 *Area Clearance2* added to the feature tree

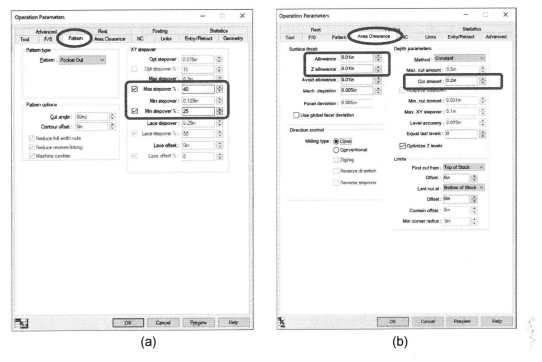

Figure 7.17 Entering machining parameters, (a) *Pattern type* and *XY stepover* under the *Pattern* tab, and (b) *Cut amount* under the *Area Clearance* tab

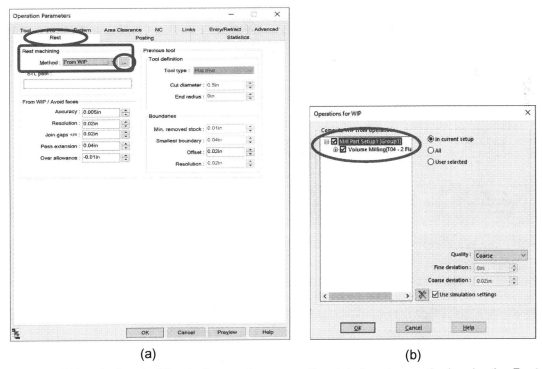

Figure 7.18 Linking the local milling to the previous operation, (a) choosing method under the *Rest* tab, and (b) picking *Rough Mill* in the *Operations for WIP* dialog box

Rename *Area Clearance2* as *Local Milling*. Right click *Local Milling* and choose *Generate Toolpath* to generate a toolpath like that shown in Figure 7.3(b).

Simulate the toolpaths by clicking the *Simulate Toolpath* button above the graphics area. Click the *Run* button ▶ to simulate the toolpath. The material removal simulation of the two operations appear in the graphics area, similar to that of Figure 7.13(b).

7.5   Adding a Multiaxis Surface Milling Operation

We now add a multiaxis surface milling operation to polish the cylindrical surface of the machined part using the same 0.5in. ball-nose cutter as that of the local milling operation.

Right click *Mill Part Setup1* and choose *Multiaxis Mill Operations > Multiaxis Mill*.

The *New Operation: Multiaxis Mill* dialog box—Figure 7.19(a)—appears. Under the *Tool* tab, select *T08 – 0.5 Ball Nose*. Click the *Features* tab, select *Multi Surface Feature1 [Area Clearance, Pattern Project]*, and click the checkmark ✔ to accept the new operation; see Figure 7.19(b).

A new operation, *Multiaxis Mill1*, is now listed in the CAMWorks feature tree tab ▦ (see Figure 7.20) and the *Operation Parameters* dialog box appears.

In the *Operation Parameters* dialog box, a 0.5in. ball-nose cutter (*ID: 115, 1/2 CRB 4FL BM 1 LOC*) is listed.

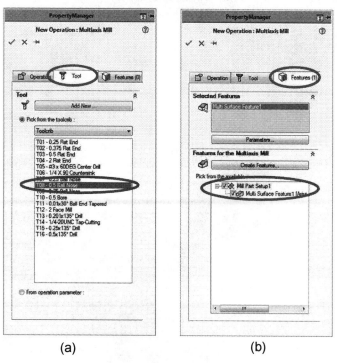

      (a)   (b)

Figure 7.19 The *New Operation: Multiaxis Mill* dialog box

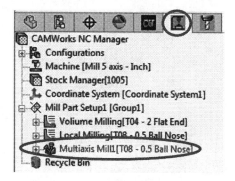

Figure 7.20 *Multiaxis Mill1* added to the feature tree

Click the *Pattern* tab, as shown in Figure 7.21(a), to review the pattern type and stepover. Change the pattern to *Flowline Between Curves* and select *Zigzag* for *Pattern* under *Direction*—circled in Figure 7.21(a). Click *Upper*; the *Curve Wizard*, as shown in Figure 7.21(b), appears. Pick the front Bézier curve of the cylindrical surface; see Figure 7.21(c) in the graphics area. Notice that the point at the left end of the curve is highlighted, indicating the start point of the curve. Click the checkmark ✔ to accept the curve.

Click *Lower* and pick the Bézier curve at the rear side of the cylindrical surface in the graphics area. Notice that the point at the left end of the curve is highlighted, indicating the start point of the curve. The start points of the curves are consistent, which leads to a valid toolpath. Choose S*tart and End At Exact Surface Edge* for *Method* under *Limits*. Change the *Max. scallop* to 0.01 in.

Click the *Axis Control* tab (Figure 7.22). Make sure *5 Axis* is chosen and select *Normal to Surface* for *Tool axis*.

Click the *Preview* button to show the toolpath like that of Figure 7.3(c). We accept the toolpath. Click the close button ✕ at the top right corner of the *Operation Parameters* dialog box and click *OK* to accept the changes and regenerate the toolpath.

Simulate the toolpaths by clicking the *Simulate Toolpath* button ⬡ above the graphics area. Click the *Run* button ▶ to simulate the toolpath. The material removal simulation of the operations appears in the graphics area, similar to that of Figure 7.13(c). The machined surface looks good. We rename the operation as *Multiaxis Surface Milling*.

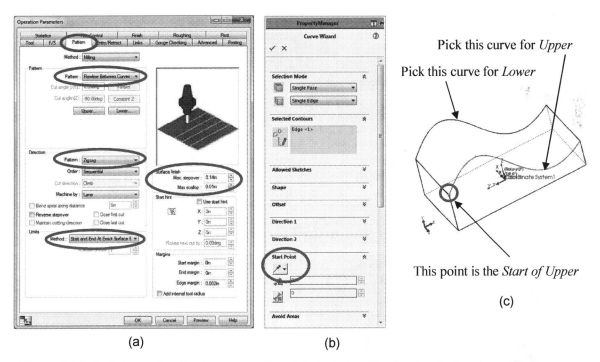

(a)   (b)

Figure 7.21 Defining a *Flowline* type operation, (a) choosing *Flowline* for *Pattern* type, (b) the curve wizard showing single edge (in this case, a curve) picked, and options of selecting start point, and (c) picking curves

7.6   Tool Gouging

Next, we replace the tool with a 2in. flat-end cutter for the surface milling operation, which is certainly inadequate. We make this change to illustrate the gouge checking capability of CAMWorks.

Right click *Multiaxis Surface Milling* and choose *Edit Definition*. Choose the 2in. flat-end mill under *Mill Tool* tab. Change the *Max stepover* to 1in. under the *Pattern* tab. Regenerate the toolpath.

The toolpath is shown in Figure 7.23(a), and the material removal simulation of the combined three operations shown in Figure 7.23(b) seems to be fine. However, after taking a closer look, for example, rotate the view similar to that of Figure 7.23(c), choose *Stock Wireframe Display* and *Target Part No Display*, from the *Toolpath Simulation* toolbox; the cutter gouges into the material at the concave area of large curvature, making excessive over cut.

Gouging is highly undesirable. How do we eliminate or avoid gouging in toolpath generation?

Right click *Multiaxis Surface Milling* and choose *Edit Definition*. Choose the *Gouge Checking* tab (see Figure 7.24), select *Group1*, apply gouge checking to *Holder*, *Shank*, *Non-cutting portion*, and *Flute*. Choose *Move Tool Away* for *Strategy* and *Along Surface Normal* for *Retract tool*. Click *OK* and regenerate the toolpath.

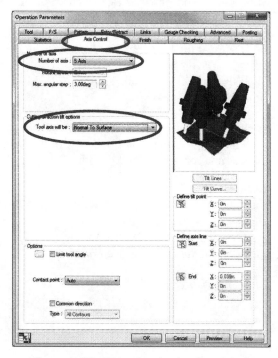

Figure 7.22 Selecting tool axis control

The regenerated toolpath is shown in Figure 7.25(a), showing that the tool is lifted up from the cylindrical surface at the concave area, where gouges occurred previously. The material removal simulation shown in Figure 7.25(b) indicates that material at the concave area is mostly uncut due to the lift of tool in the area to avoid gouging. A closer look at material removal simulation verifies that the tool is lifted up in the area to avoid gouging; see Figure 7.25(c).

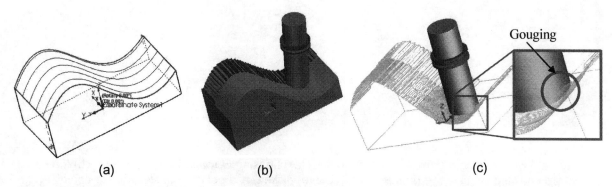

| (a) | (b) | (c) |

Figure 7.23 Illustration of tool gouging, (a) toolpath of the multiaxis surface milling operation using a 2in. flat-end cutter, (b) material removal simulation, and (c) tool gouging seen in a zoom-in view

Restore the previous surface milling operation by turning the gouge checking off, bringing the 0.5in. ball-nose cutter back, entering the *Max. scallop* to 0.01 in., and regenerate the toolpath.

7.7 Machine Simulation

Machine Simulation in CAMWorks offers a more realistic setup in simulating machining operations. For example, a sample machine simulator called *Mill_Tutorial* (among others offered by CAMWorks), provides a setup of tools with a tool holder and stock mounted on a tilted rotary table, as shown in Figure 7.1(b).

Machine Simulation also detects tool collisions when they happen. Note that you may add your own virtual CNC machines, a virtual replica of the physical machine at your machine shop, to CAMWorks that allows you to carry out a more realistic machining simulation.

You may also add associated post processors to CAMWorks that support generating G-codes compatible with the respective CNC machines. More about adding virtual machines to CAMWorks will be discussed in Lesson 10.

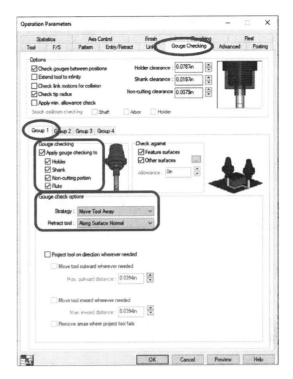

Figure 7.24 Defining gouge checking in the *Operation Parameters* dialog box

To bring up the Machine Simulation window, you may right click *Mill Part Setup* under CAMWorks operation tree ▧ , and choose *Machine Simulation* and then *Legacy*.

The basic layout of the *Machine Simulation* window is shown in Figure 7.26. In the graphics area, the cutter, tool holder, tilt table, rotary table, and the stock together with a machine coordinate system XYZ are displayed. To the right, the three machining operations and corresponding G-codes are listed in the upper and lower areas of *Move List*, respectively.

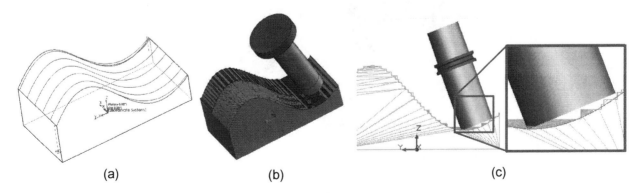

| (a) | (b) | (c) |

Figure 7.25 Toolpath avoiding gouging, (a) toolpath of the modified multiaxis surface milling operation, (b) material removal simulation, and (c) tool lifted up from the cylindrical surface to avoid gouging

On top, the default simulator, *Mill_Tutorial*, is selected, with offsets in X-, Y-, and Z-directions all set to be zero. Below are buttons that control the machining simulation run. Under these buttons are four tabs, *FILE*, *SIMULATION*, *VERIFICATION*, and *VIEW*, supporting respective uses of the machine simulation capability. *SIMULATION* tab is chosen as default. Under the *SIMULATION* tab, there are *Simulation* options that offer setups to display the desired simulation, such as material removal. Next to the *Simulation* options are buttons that provide options for *Control*, *Simulation Run Speed*, and *Views*.

Before we start, click the *Select* button ▣ next to the *Machine* on top of the *Machine Simulation* window (Figure 7.26) to bring up the *Select Point* dialog box (Figure 7.27). We are about to select the origin of the work coordinate system on the part that determines its position and orientation with respect to the top face of the rotary table. Pick the coordinate system, *Coordinate System1*, in the graphics window (see Figure 7.28), to choose the origin of *Coordinate System1* as the origin of the work coordinate system. And then, click the *Update* button on top. The origin of *Coordinate System1* (at the center of the bottom face of the stock) coincides with the center point at the top face of the rotary table (see Figure 7.26). Also, the offsets in X-, Y-, and Z-directions are all zero.

To verify if a correct coordinate system has been chosen for the Machine Simulation, you may choose *TableTable* as the machine by pulling down the *Machine* selection and choosing it (see Figure 7.29). Click the *Update* button. In the graphics area, a work coordinate system appears at the center of the bottom face of the stock with its X-, Y-, and Z-axes parallel to those of the virtual machine shown at the lower left corner of Figure 7.30.

Now, we choose *Mill_Tutorial* as the machine and click the *Update* button and run the simulation.

Click the *Run* button ▶ to start the simulation. The G-code of the corresponding tool motion is displayed in the lower portion of the *Move List* area.

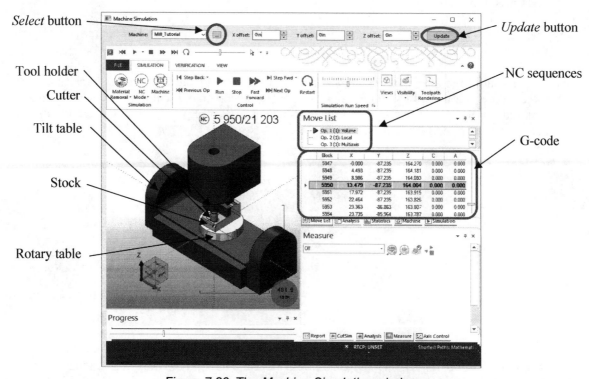

Figure 7.26 The *Machine Simulation* window

The simulation runs to the end for all three operations without any issue. There is no collision encountered. Next, we create a scenario that shows collision detection capability of the Machine Simulation.

Close the *Machine Simulation* window. Suppress the first two operations by right clicking *Volume Milling* (and *Local Milling*) and choosing *Suppress* under the CAMWorks operation tree 🖹. Rerun the machine simulation. The simulation starts from the third operation, *Multiaxis Surface Milling*, directly since the first two operations were suppressed.

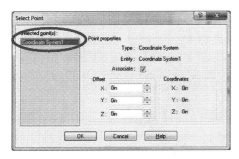

Figure 7.27 The *Select Point* dialog box

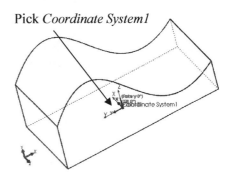

Pick *Coordinate System1*

Figure 7.28 Picking the coordinate system, *Coordinate System1*

Figure 7.29 Choosing *TableTable* for machine

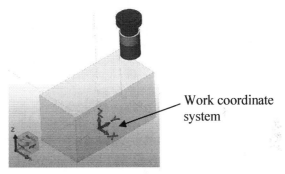

Work coordinate system

Figure 7.30 The work coordinate system displayed

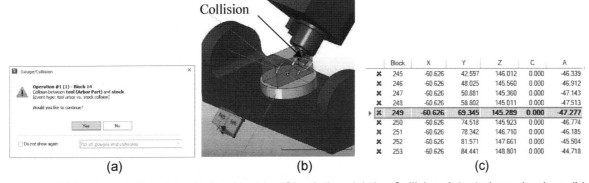

	Block	X	Y	Z	C	A
✖	245	-60.626	42.597	146.012	0.000	-46.339
✖	246	-60.626	48.025	145.560	0.000	-46.912
✖	247	-60.626	50.881	145.360	0.000	-47.143
✖	248	-60.626	58.802	145.011	0.000	-47.513
✖	249	-60.626	69.345	145.289	0.000	-47.277
✖	250	-60.626	74.518	145.923	0.000	-46.774
✖	251	-60.626	78.342	146.710	0.000	-46.185
✖	252	-60.626	81.571	147.661	0.000	-45.504
✖	253	-60.626	84.441	148.801	0.000	-44.718

(a)   (b)   (c)

Figure 7.31 The collision detection in *Machine Simulation*, (a) the *Collision detected* warning box, (b) collision area highlighted, and (c) the symbol ✖ appearing in front of problematic NC blocks

Without the volume milling and local milling, a collision between the tool holder and the stock is detected immediately after the simulation starts. Click *Yes to All* in the warning box, as shown in Figure 7.31(a), to allow the simulation to continue. The collision area is highlighted in red—Figure 7.31(b)—and a symbol ✖ appears in the problematic NC blocks in the G-code where collisions are detected; see Figure 7.31(c).

Close the *Machine Simulation* window. Unsuppress the first two operations by right clicking *Volume Milling* (and *Local Milling*) and choosing *Unsuppress* under the CAMWorks operation tree 🖫.

We have completed the exercise. You may save the model for future reference. Next, we revisit the freeform surface example of Lesson 4.

We create a multiaxis surface milling operation using a 5-axis mill in place of the *Surface Milling* operation in hopes of eliminating or reducing the noticeable material remaining at numerous areas; for example, near the middle of the lower edge of the freeform surface, as indicated in Figure 4.35 of Lesson 4.

7.8   Revisiting the Freeform Surface Example of Lesson 4

We now open the example file saved from Lesson 4, in which three operations were generated, *Volume Milling*, *Local Milling*, and *Surface Milling*, using a 3-axis mill.

Under CAMWorks operation tree 🖫 , right click *Mill Part Setup1*, and choose *Multiaxis Mill Operation > Multiaxis Mill*.

We follow the same steps outlined in Section 7.5 to create a multiaxis surface mill operation. That is, we pick the freeform surface as the machinable feature, and choose a 0.5in. ball-nose cutter (T08).

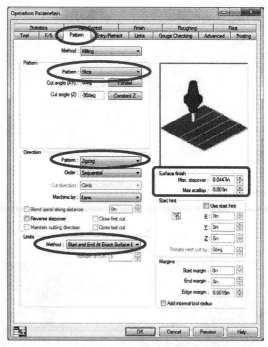

Under the *Pattern* tab of the *Operation Parameters* dialog box (see Figure 7.32), we choose *Slice* for *Pattern*, choose *Zigzag* for direction pattern, choose *Start and End At Exact Surface Edges* for *Method*, and enter *0.001* for *Max scallop*.

We define tool axis control using the exact same options as those of Figure 7.22; i.e., 5 axis and normal to surface.

The resulting toolpath of the multiaxis surface milling is shown in Figure 7.33(a). Because of the options we chose, *Start and End At Exact Surface Edges* for *Method*, and the fact that a 5-axis mill is employed, the toolpaths show that two passes of the toolpath stay right on the two respective boundary edges of the freeform surface. As a result, the noticeable material near the boundary edges seen in Lesson 4 will be removed.

Figure 7.32 Defining a 5-axis surface mill operation under the *Pattern* tab

The material removal simulation of the combined three operations is shown in Figure 7.33(b) and Figure 7.33(c) (section view). The noticeable amount of material uncut near the lower edge of the freeform surface as seen in Lesson 4 is completely eliminated, as predicted. Same is true near the top edge. Note that the areas of purple color appearing on the freeform surface is simply due to visualization. Tilting the view slightly leads to a machined surface without purple areas (see Figure 7.34).

We have completed revisiting this example. You may save your model for future reference.

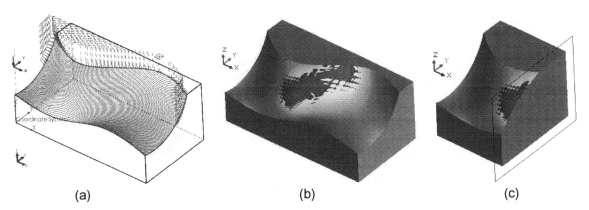

| (a) | (b) | (c) |

Figure 7.33 Toolpath and material removal simulation of the freeform surface example from Lesson 4, (a) toolpath of the multiaxis surface milling operation, (b) material removal simulation of the combined machining operations, and (c) section view at offset 3.3in. in YZ plane

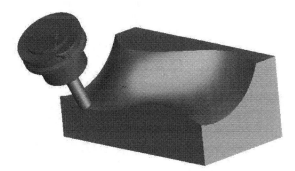

Figure 7.34 The machined surface at the end of the material removal simulation of combined toolpaths in a slightly tilted view

7.9   Exercises

Problem 7.1. In this exercise, we rework the freeform surface example of Problem 4.1.

(a)   Replace the *Pattern Project* operation by creating a multiaxis surface milling operation using a 5-axis mill. Would a 5-axis mill offer a better toolpath than that of *Pattern Project*, especially near the lower edge of the freeform surface where notable material remained uncut?

(b)   Adjust the allowance parameters of the *Volume Milling* and *Local Milling* operations to understand the effect of the parameters to the quality of the machined surface. For example, set allowances to zero for both operations. Generate toolpath and carry out material removal simulation for all three operations combined. Can you identify the undesired tool marks on the freeform surface?

(c)   Simulate the machining operations using the *Machine Simulation* capability, similar to that of Figure 7.35. Is any tool collision detected?

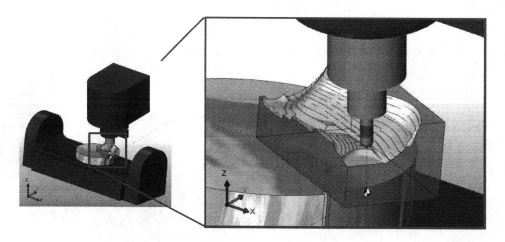

Figure 7.35 Machine simulation of Problem 7.1

Problem 7.2. We explore more options in the *Operation Parameters* dialog box for multiaxis surface milling. Use the example you just created after going through this lesson (or the example file *Freeform Surface with Toolpath* downloaded) as a reference, and choose the following options. Generate toolpaths and carry out material removal simulations. In each option, summarize the characteristics of the toolpaths, the advantages, and disadvantages.

(a)   Click the *Pattern* tab, as shown in Figure 7.21(a), and select *Zig* (instead of *Zigzag*) for *Pattern* under *Direction*—circled in Figure 7.21(a).

(b)   Click the *Entry/Retract* tab and choose *Use Leadin* (and *Leadout*) for *Method*. Also, adjust the length/width and arc to see the change in toolpath.

(c)   Click the *Axis Control* tab and choose *Tilted Relative to Cutting Direction* for Tool axis.

Lesson 8:   Turning a Stepped Bar

8.1   Overview of the Lesson

We discuss turning operations in Lessons 8 and 9. In Lesson 8, we use a simple stepped bar example to learn basic capabilities in creating turning operations and understanding G-code post-processed by CAMWorks. In Lesson 9, we machine a similar example with more turning features to gain a broader understanding of the turning capabilities offered by CAMWorks.

This current lesson should provide you with a quick run-through in creating turning operations using CAMWorks. You will learn a complete process in using CAMWorks to create turning operations from the beginning all the way to the post process that generates G-code. We use a lathe of single turret to machine the simple stepped bar shown in Figure 8.1(a) from a round stock clamped into a three-jar chuck, as shown in Figure 8.1(b). We create an outer profile turning operation (called outer diameter or OD turning in CAMWorks) and a cut off operation to remove the part from the stock. The machined part at the end of the material removal simulation is shown in Figure 8.1(c). This lesson is intentionally made simple. We stay with default options and parameters for most of the lesson.

Similar to the milling operation lessons, we follow the general steps shown in Figure 1.1 of Lesson 1 to turn the stepped bar. We will start by (1) opening the stepped bar design model; (2) defining machine setup, in which we choose a single turret lathe, select a tool crib, pick a post processor, and choose a fixture coordinate system; (3) creating a stock, in this case, a cylindrical bar stock enclosing the design model; (4) defining machinable features manually and later using automatic feature recognition (AFR) in Lesson 9; (5) generating operation plans and toolpaths; and (6) checking results and reviewing material removal simulation. In addition, we will go over a post process to create G-code of selected operations, including turn finish and cut off. We will take a closer look at the G-code generated to gain a better understanding of the turning operations.

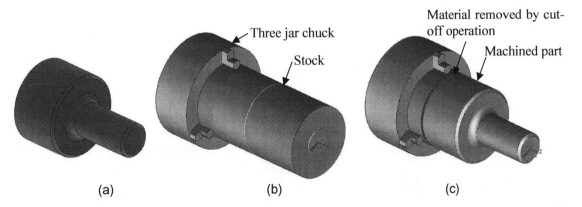

(a)   (b)   (c)

Figure 8.1 The stepped bar example, (a) the design model, (b) bar stock clamped to a chuck, and (c) the material removal simulation

After completing this lesson, you should be able to carry out machining simulation for similar problems following the same procedures and be ready to move onto Lesson 9.

8.2   The Stepped Bar Example

The stepped bar shown in Figure 8.2(a) has a bounding cylinder of size ϕ4.25in.×6.5in. The unit system chosen is IPS. There is one revolve feature created by revolving the sketch shown in Figure 8.2(b). In addition, there is a chamfer of 0.15in.—defined at two boundary edges shown in Figure 8.2(a)—and a fillet (0.375in. in radius) feature, plus a point and a coordinate system. These features are listed in the feature tree shown in Figure 8.3. The coordinate system (*Coordinate System1*) and the point (*Point1*) are chosen as the fixture coordinate system and the part setup origin, respectively, for turning operations. When you open the solid model *Stepped bar.SLDPRT*, you should see the solid and the reference features listed in the feature tree like those shown in Figure 8.3.

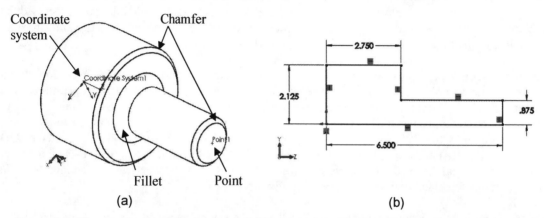

Figure 8.2 The stepped bar design model, (a) solid and reference features, and (b) sketch of the revolved solid feature

A bar stock of size ϕ4.25in.×7.75in., made of low carbon alloy steel (1005), as shown in Figure 8.4(a), is chosen for the turning operations. The front end of the stock coincides with the front face of the part. The rear end of the stock is clamped in the three-jar chuck with an extra length of 1.25in. to be cut off at the end of the turning operation. Since the front end face of the part coincides with that of the stock, there is no need to define a face turning operation.

Note that a part setup origin is defined at the center of the front end face of the stock (coinciding with *Point1* defined in part), which defines the G-code program zero location.

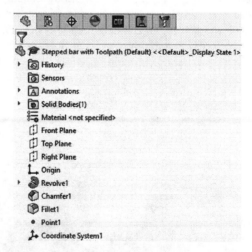

Figure 8.3 Entities listed in the feature tree

We will create two machinable features, an OD (outer diameter) feature that includes the outer shape of the part from the front end face to the cut off face, and a cut off feature to cut off the part from the stock. In general, an OD machinable feature excludes the shape of any groove features (which are not present in this simple example).

We create all machinable features manually. In the next lesson, we will use the automatic feature recognition (AFR) capability to recognize machinable features from the part solid model. In this lesson, we follow the recommendations of the technology database (TechDB™) for choosing machining options, cutters, and cutting parameters.

The toolpaths of the turning operations for machining the OD feature, consisting of *Turn Rough* and *Turn Finish*, are shown in Figure 8.5(a) and Figure 8.5(b), respectively. The turn rough toolpath moves the cutter along the outer profile of the part in numerous passes. This is because the cut amount is chosen as 0.1in. and the overall depth to cut is 1.25in. The turn finish toolpath moves the cutter along the part outer boundary in one pass, similar to the contour mill operation in milling. The cut off toolpath simply moves the cutter along the negative X direction to separate the part from the stock, as shown in Figure 8.5(c).

You may open the example file with toolpath created (filename: *Stepped bar with toolpath.SLDPRT*) to preview its toolpaths and machining simulation.

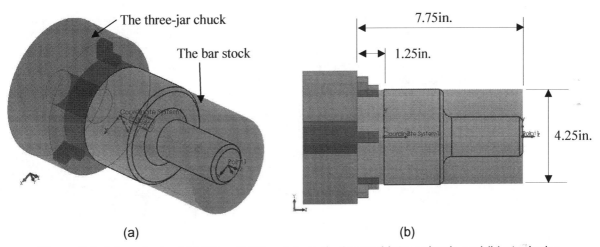

(a)   (b)

Figure 8.4 A bar stock of ϕ4.25in.×7.75in., (a) stock clamped into a chuck, and (b) stock size

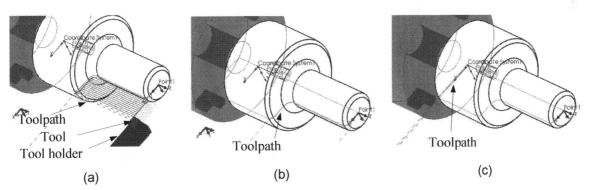

(a)   (b)   (c)

Figure 8.5 The toolpaths of the turning operations, (a) *Turn Rough*, (b) *Turn Finish*, and (c) *Cut Off*

8.3 Using CAMWorks

Open SOLIDWORKS Part

Open the stepped bar design model (filename: *Stepped bar.SLDPRT*) downloaded from the publisher's website. This solid model, as shown in Figure 8.3, consists of three solid features (revolve, chamfer, and fillet), a point, and a coordinate system. As soon as you open the model, you may want to check that the IPS unit system is chosen and increase the decimals from the default 2 to 4 digits.

Select NC Machine

Click the CAMWorks feature tree tab ▦. Right click *Machine* and select *Edit Definition*. In the *Machine* dialog box (Figure 8.6), *Mill-inch* is selected under *Machine* tab. We choose *Turn Single Turret - Inch* from the list of *Applicable machines* box and click *Select*.

Choose *Tool Crib* tab and select *Tool Crib 2 Rear* under *Available tool cribs* (Figure 8.7), and then click *Select*. Choose the *Post Processor* tab; a post processor called *T2AXIS-TUTORIAL* is selected (Figure 8.8). This is a generic post processor of the 2-axis lathe that comes with CAMWorks. There are more post processors that come with CAMWorks; they are located in *C:\ProgramData\SOLIDWORKS\CAMWorks 2023\Posts* folder in your computer.

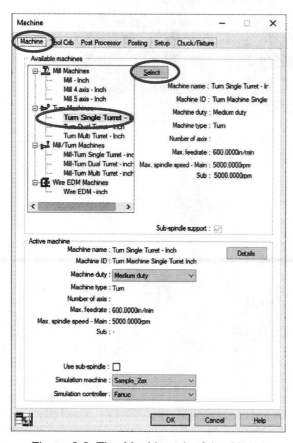

Figure 8.6 The *Machine* tab of the *Machine* dialog box

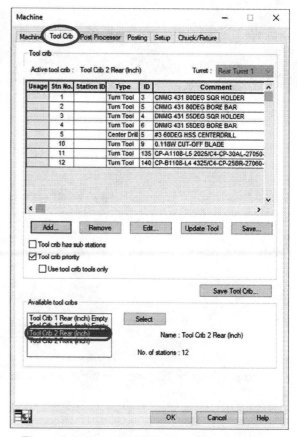

Figure 8.7 The *Tool Crib* tab of the *Machine* dialog box

Note that in practice you will need to identify a suitable post processor that produces G-code compatible with the CNC lathes at the shop floor. We will not make any change in the post processor selection for this lesson. Then, click the *Setup* tab, and click *User Defined* under *Main spindle* and choose *Coordinate System1* for *Coordinate system* (see Figure 8.9). Click *OK* to accept the selections and close the dialog box.

Create Stock

From CAMWorks feature tree [CW], right click *Stock Manager* and choose *Edit Definition*. The *Stock Manager* dialog box appears (Figure 8.10), in which a default stock size appears under *Bar stock parameters*, including outer diameter ▣ (4.25in.), inner diameter ▣ (0in.), overall length ▣ (6.5in.), and back of stock absolute ▣ (0in.) that defines length of the stock outside the part.

We choose *Low Carbon Alloy Steel 1005* for stock material. We will increase the overall length of the stock to 7.75in. and enter the length of the stock outside the part to be –1.25in., as shown in Figure 8.10. Accept the revised stock by clicking the checkmark ✔ at the top left corner. The bar stock should appear in the graphics area similar to that of Figure 8.4.

Turn Setup and Machinable Feature

We create two machinable features manually, OD and cut off. We first create a turn setup and then insert new turn features under the CAMWorks feature tree.

Figure 8.8 The *Post Processor* tab of the *Machine* dialog box

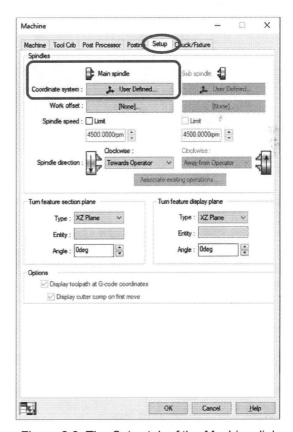

Figure 8.9 The *Setup* tab of the *Machine* dialog box

Under the CAMWorks feature tree tab , right click *Stock Manager* and choose *Turn Setup*. The *Turn Setup* dialog box appears (Figure 8.11). In the graphics area, a coordinate system of the part setup appears at the rear end face of the part (circled in Figure 8.12), coinciding with the fixture coordinate system, *Coordinate System1*. This coordinate system indicates that the Z-axis aligns with the spindle direction, which is adequate.

We will relocate the part setup origin to the front end face of the stock later. For the time being, click the checkmark ✔ in the *Turn Setup* dialog box to accept the definition. A *Turn Setup1* is now listed in the CAMWorks feature tree .

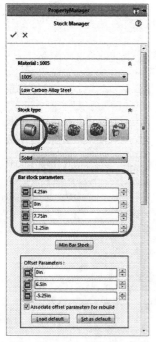

Figure 8.10 The *Stock Manager* dialog box

Figure 8.11 The *Turn Setup* dialog box

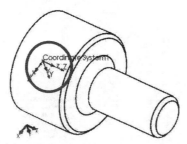

Figure 8.12 The turning coordinate system

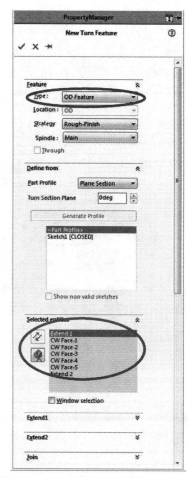

Figure 8.13 The *New Turn Feature* dialog box

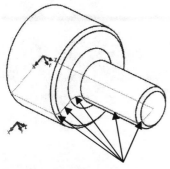

Pick these five edges

Figure 8.14 Picking these five edges to define an OD feature

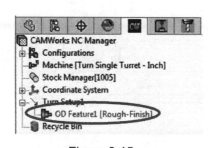

Figure 8.15

Now we define machinable features. Under CAMWorks feature tree tab ▣, right click *Turn Setup1* and choose *Turn Feature*. In the *New Turn Feature* dialog box (Figure 8.13), choose *OD Feature* for *Type*, and pick the five edges that define the boundary profile of the OD feature in the graphics area, as shown in Figure 8.14. Click the checkmark ✓ in the *New Turn Feature* dialog box to accept the definition.

An *OD Feature1* node is now listed in the CAMWorks feature tree ▣ in magenta color (see Figure 8.15).

We follow the same steps to create a cut off machinable feature; i.e., right click *Turn Setup1* and choose *Turn Feature*. In the *New Turn Feature* dialog box, choose *CutOff Feature* for *Type*, and pick the edge at the rear end face of the part in the graphics area, as shown in Figure 8.16. Click the checkmark ✓ in the *New Turn Feature* dialog box to accept the definition.

A *CutOff Feature1* node is now listed in the CAMWorks feature tree ▣ in magenta color.

Generate Operation Plan and Toolpath

Right click *Turn Setup1* and choose *Generate Operation Plan* (or click the *Generate Operation Plan* button ▣ above the graphics area).

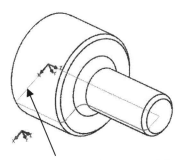

Pick this edge at the rear end face

Figure 8.16 Picking this edge
to define a cutoff feature

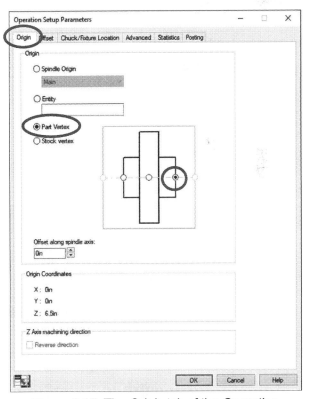

Figure 8.17 The *Origin* tab of the *Operation
Setup Parameters* dialog box

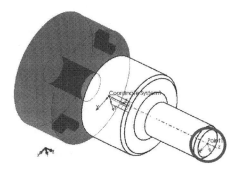

Figure 8.18 The part setup origin
relocated to the center of front end face

Three operations, *Turn Rough1*, *Turn Finish1*, and *Cut Off1*, are listed in CAMWorks operation tree [K] in magenta color. Right click *Turn Setup1* and choose *Generate Toolpath* (or click the *Generate Toolpath* button [icon] above the graphics area). Turning toolpaths will be generated like those shown in Figure 8.5.

Relocate Part Setup Origin

A part setup origin has been chosen by CAMWorks to be coincident with the fixture coordinate system, *Coordinate System1* (see Figure 8.12). We will move the part setup origin to *Point1* (located at the center of the front end face of the part), which is more practical since this point is easier to access in a stock clamped into a chuck on a lathe.

Under the CAMWorks operation tree tab [K] , right click *Turn Setup1* and choose *Edit Definition*.

In the *Operation Setup Parameters* dialog box, choose *Origin* tab, and select *Part Vertex*. The vertex at the right end of the part in the sketch (circled in Figure 8.17) is selected, which matches our intent. The part setup origin moved to the center of the front end face (see Figure 8.18) as desired. Click *OK* in the *Operation Setup Parameters* dialog box to accept the change. Click *Yes* to the question in the warning box: *The origin or chuck/fixture location/avoidance parameters has changed, toolpaths need to be recalculated. Regenerate toolpaths now?*

Next we review the tool and key machining parameters of the first turning operation, *Turn Rough1*.

Review the Operations

Under the CAMWorks operation tree tab [K] , right click *Turn Rough1* and choose *Edit Definition*. In the *Operation Parameters* dialog box, choose *Diamond Insert* tab; a diamond insert of 0.0157×80° (radius 0.0157in., angle: 80°, and inscribed circle 0.5in., ID:1, CNMG 431) appears.

Change the insert radius to 0.02 (circled in Figure 8.19). Click the *Holder* tab (see Figure 8.20). A standard holder of shank width 0.75in. and length 4in. (Holder ID:1, RH 80DEG SQR HOLDER) has been chosen. Also, the *Down left* is chosen for *Orientation*, which defines the orientation of the cutter suitable to turn the OD feature.

Choose the *Turn Rough* tab of the *Operation Parameters* dialog box. In the *Profile parameters* area, the *First cut amt.* and *Max cut amt.* are set to *0.1in*, and *Final cut amt.* is *0.025in.*, as shown in Figure 8.21. These parameters define the distance of the tool movement along the negative X-direction, similar to the depth of cut in volume milling. Click *OK* to accept the selections and close the dialog box. You may review other operations following the same steps.

Note that we also change the radius of the diamond insert to 0.02 for the *Turn Finish1* operation as well. This is to simplify a bit for reviewing the toolpath next.

Simulate Toolpath

Right click *Turn Setup1* and choose *Simulate Toolpath* (or click the *Simulate Toolpath* button [icon] above the graphics area) to the simulate toolpath. The material removal simulation appears similar to that of Figure 8.1(c).

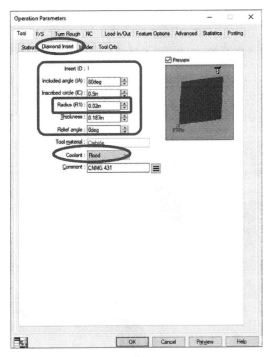

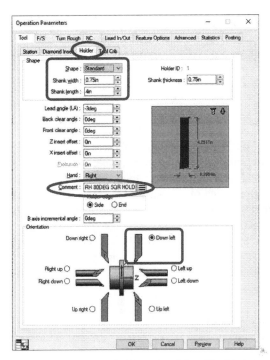

Figure 8.19 The *Diamond Insert* tab of the *Tool* tab in the *Operation Parameters* dialog box

Figure 8.20 The *Holder* tab of the *Tool* tab in the *Operation Parameters* dialog box

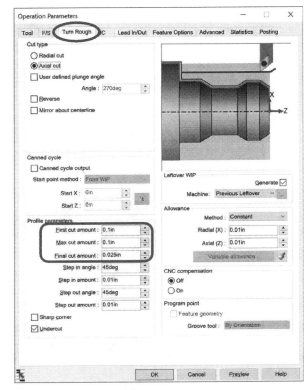

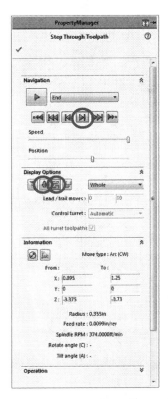

Figure 8.21 The *Turn Rough* tab in the *Operation Parameters* dialog box

Figure 8.22 The *Step Through Toolpath* dialog box

Step Through Toolpath

Now we take a closer look at the toolpath using the *Step Through Toolpath* capability to better understand the turning operations generated.

We pick two operations for a closer look, *Turn Finish* and *Cut Off*, since both are simple and easier to understand.

In *Turn Finish* operation, the tool moves along the five edges of the OD machinable feature. In *Cut Off* operation, the tool cuts along the straight edge at the rear end face of the part. Reviewing the toolpath will help us understand the G-code to be discussed shortly.

Right click *Turn Finish1* and choose *Step Thru Toolpath*. The *Step Through Toolpath* dialog box appears (Figure 8.22).

Under *Information*, CAMWorks shows the tool movement from the current to the next step in X, Y, and Z coordinates, with the feedrate, spindle speed, and other machining information.

Click the *Step* button ▶ at the center of the *Step Through Toolpath* dialog box (circled in Figure 8.22) to step through the toolpath. You may want to turn on *Show toolpath points* and *Tool Holder Shaded Display* (circled in Figure 8.22) to see the toolpath display similar to that of Figure 8.23.

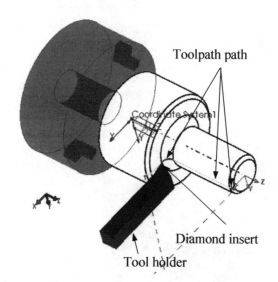

Figure 8.23 Stepping through the toolpath of *Turn Finish1* operation

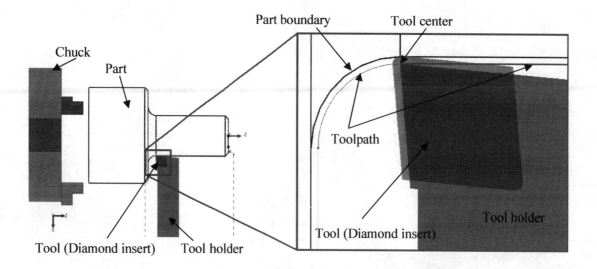

Figure 8.24 Toolpath of *Turn Finish1* operation, a closer look

A closer look at the toolpath of *Turn Finish1* operation (see Figure 8.24) indicates that the toolpath follows the trace of the tool center (i.e., the center of the corner radius of the insert). As a result, in general, the toolpath offsets an amount of insert radius from the part boundary. The XZ coordinates of the six characteristic points of the part boundary, A to F shown in Figure 8.25, referring to the part setup origin are listed in Table 8.1. The toolpath offset by an amount of the insert radius, 0.02in., is especially clear at points B, C, D and E. Note that if you see different results, you may want to check if the insert radius of *Turn Finish1* operation was changed to 0.02.

Now we follow the same steps to review the toolpath of the *Cut Off* operation.

Right click *Cut Off1* and choose *Step Thru Toolpath*. Note that the Z coordinates shown in the *Step Through Toolpath* dialog box (Figure 8.26) are –6.559, while the rear end face of the part is 6.5in. to the left of the part setup origin (again, located at the center of the front end face of the part). This is because a groove cutter of 0.118in. in width was selected by CAMWorks for the operation. The right edge of the tool is in contact with the part boundary, as seen in Figure 8.27. Therefore, the toolpath is offset an amount of half cutter width to the left of the part boundary. That is, the Z-coordinates of the toolpath are –6.5–0.118/2 = –6.559, as expected.

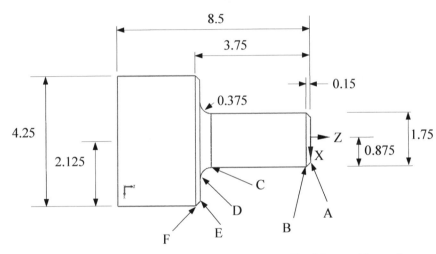

Figure 8.25 The six characteristic points (A to F) of the part boundary

Table 8.1 The XZ Coordinates of the Six Characteristic Points A to F

Point	X and Z Coordinates of Part Boundary	X and Z Coordinates of Toolpath	Offsets in X and Z Coordinates
A	X = 0.725 Z = 0	X = 0.7391 Z = 0.0141	$\Delta X = 0.0141$ $\Delta Z = 0.0141$
B	X = 0.875 Z = –0.15	X = 0.895 Z = –0.15	$\Delta X = 0.02$ $\Delta Z = 0$
C	X = 0.875 Z = –3.375	X = 0.895 Z = –3.375	$\Delta X = 0.02$ $\Delta Z = 0$
D	X = 1.25 Z = –3.75	X = 1.25 Z = –3.73	$\Delta X = 0$ $\Delta Z = 0.02$
E	X = 1.975 Z = –3.75	X = 1.975 Z = –3.73	$\Delta X = 0$ $\Delta Z = 0.02$
F	X = 2.125 Z = –3.9	X = 2.1391 Z = –3.8859	$\Delta X = 0.0141$ $\Delta Z = 0.0141$

8.4   Reviewing the G-code of Turning Operations

We are now ready to generate and review G-code for *Turn Finish1* and *Cut Off1* operations. Note that we choose to output the G-code at tool nose center so that the code can be verified with the selected CL data discussed above.

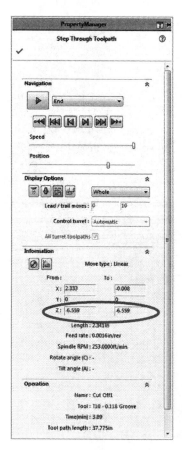

Figure 8.26 The *Step Through Toolpath* dialog box

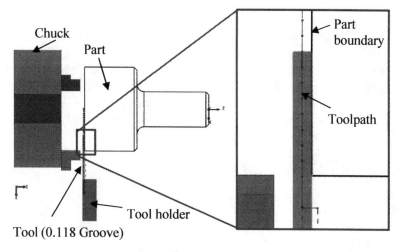

Figure 8.27 Toolpath of *Cut Off*, a closer look

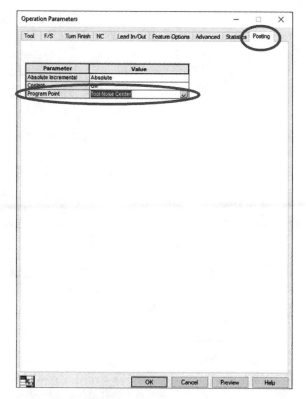

Figure 8.28 The *Posting* tab of the *Operation Parameters* dialog box

We right click *Turn Finish1* and choose *Edit Definition*. In the *Operation Parameters* dialog box, choose *Posting* tab, and select *Tool Nose Center* for *Program Point* (circled in Figure 8.28). Click *OK* to accept the selection.

Right click *Turn Finish1* and choose *Post Process*. In the *Post Output File* dialog box (Figure 8.29), choose a proper file folder, enter a file name (for example, *Turn finish.txt*). The *Post Process* dialog box appears (Figure 8.30).

In the *Post Process* dialog box, click the *Play* button ▶ (circled in Figure 8.30) to create G-code (.txt file).

Open the *Turn finish.txt* file from the folder using *Word* or *Word Pad*; see the file contents shown in Figure 8.31(a). It is shown that the NC blocks N8 and N9 moves the cutter to Point A, N11 to Point B, N12 to Point C, N13 to Point D, N14 to Point E, and N16 to Point F. Note that the X locations of the cutter are output as diametral instead of radial determined by the post processor employed.

Repeat the same for the cut off operation (choosing tool nose center). The NC block N8 shows the Z-coordinate of the toolpath; see Figure 8.31(b). Note that the Z-coordinate shown in the G-code is –6.559, which is correct.

We have now completed the lesson. You may save your model for future reference.

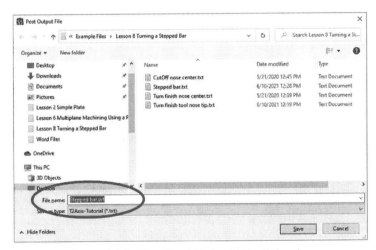

Figure 8.29 The *Post Output File* dialog box

Figure 8.30 The *Post Process* dialog box

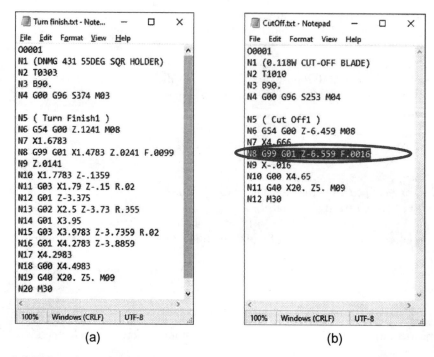

Figure 8.31 Turn G-codes, (a) *Turn Finish1* operation, and (b) *Cut Off1* operation

8.5   Exercises

Problem 8.1. Repeat the same steps in this lesson to generate turning operations, except that we use the automatic feature recognition capability by clicking the *Extract Machinable Features* button above the graphics area. Is there any redundant machinable feature extracted? Generate toolpath. Are there any noticeable differences in the toolpath of the three turning operations, turn rough, turn finish, and cut off, compared with those discussed in this lesson?

Problem 8.2. Generate NC operations to machine the part shown in Figure 8.32 from a stock of φ2.75in.×4.5in. Pick adequate tools (with justifications). Please submit the following for grading:

(a)   A summary of the turn operations, including cutting parameters and tools selected.
(b)   Screen shots of combined NC toolpaths and material removal simulations.
(c)   Is there any material remaining uncut? If so, was it a part design issue or machining issue? What can be done to improve either the part design or machining operations to correct the issue?

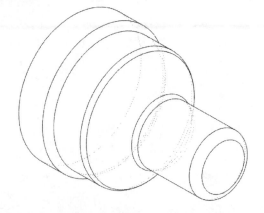

Figure 8.32 The design model of Problem 8.2

Lesson 9:   Turning a Stub Shaft

9.1   Overview of the Lesson

In Lesson 9, we machine a stub shaft, which is similar to that of Lesson 8 but with more turning features, including face, groove, thread, and holes at both ends. Since the two holes are located at the front and rear ends, respectively, there will be two turn setups that support turning operations to machine the features from both ends of the bar stock.

Lesson 9 offers a more in-depth discussion in creating turning operations using CAMWorks. You will learn to use automatic feature recognition (AFR) to extract machinable features and make necessary adjustments for a valid turning simulation that can be implemented physically. We will use the same machine as Lesson 8, that is, a lathe of single turret, to machine the stub shaft example shown in Figure 9.1(a) from a bar stock clamped into a three-jar chuck from its rear end, then to the front end, to turn machinable features from both ends. Note that only the chuck at the rear end is shown in Figure 9.1(b). Therefore, there are at least two turn setups required for this example. The machined part at the end of the material removal simulation of the first turn setup (*Turn Setup1*) is shown in Figure 9.1(c).

Most of the turning operations generated by CAMWorks are adequate for this example, except for the threading operation. Although we stay with default options and parameters for most of the lesson, we point out a few deficiencies that need to be corrected, including the threading operation. We will take a closer look at the threading operation by stepping through the toolpath and reviewing the G-code to gain a better understanding.

At the end we briefly introduce the procedure of using a turn-mill to machine the two side cuts and a cross hole near the mid portion of the shaft. These features will have to be taken care of by using a mill instead of a lathe. A turn-mill is suitable to machine such a part with both turn and mill machinable features.

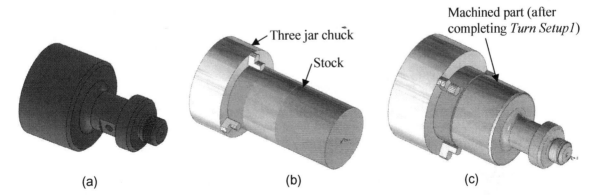

Figure 9.1 The stub shaft example, (a) the design model, (b) bar stock clamped to a chuck at the rear end, and (c) the material removal simulation (after completing *Turn Setup1*)

After completing this lesson, you should be able to carry out a turning simulation for applications of most turning features following the same procedures.

9.2   The Stub Shaft Example

Similar to the stepped bar example, the stub shaft has a bounding cylinder of size $\phi 4.25$in.$\times 6.5$in. In addition to the revolve, chamfer and fillet features, there are side cuts, a cross hole, and holes at the two ends; see Figure 9.2(a). Note that the sketch of the revolve feature shown in Figure 9.2(b) is a bit more complex than that of Lesson 8, which leads to additional machinable features, such as groove features. The sketches of the end holes and thread are shown in Figure 9.2 (c). Note that the thread pitch is 0.125in. The reference features, including a point and a coordinate system, are similar to those of Lesson 8. These features are listed in the feature tree shown in Figure 9.3. Similar to Lesson 8, the coordinate system (*Coordinate System1*) is chosen as the fixture coordinate system. Both the origin of *Coordinate System1* and *Point1* are defined as the part setup origins, respectively, for the two turn setups. When you open the solid model *Stub shaft.SLDPRT*, you should see the solid features and the reference features listed in the feature tree (see Figure 9.3).

The unit system chosen is IPS. Also, like before, we increase the decimal points from 2 to 4.

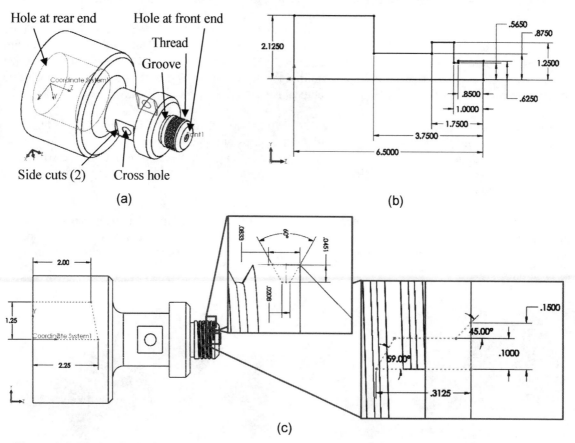

Figure 9.2 The stub shaft solid model, (a) major solid features, (b) sketch and dimensions of the revolve solid feature, and (c) sketches and dimensions of the two end holes and thread

A bar stock of size ϕ4.25in.×8in., made of low carbon alloy steel (1005), as shown in Figure 9.4(a), is chosen for the machining operations. The front end of the stock is extended by 0.25in. from the front face of the part. The rear end of the stock is fixed in the three-jar chuck with an extra length of 1.25in. to be cut off at the end of the turning operations (similar to that of Lesson 8) of the first turn setup. Due to the extra material at the front end face, a face turning operation is required.

Note that the part setup origins are defined at the center of the front end face of the stock and the origin of the coordinate system, *Coordinate System1*, for the two turn setups, respectively. The first setup cuts all features except for the hole at the rear end, which is machined by the operations of the second setup.

We first extract machinable features using the automatic feature recognition (AFR) capability. Most machinable features are extracted, including face, OD, grooves, ID (the two holes). The only feature that is not extracted is the thread. We manually create a thread machinable feature. Overall, there are eight machinable features included, as shown in Figure 9.5. In this lesson, we follow the recommendations of the technology database (TechDB™) for determining machining options, tools, and cutting parameters.

The side cuts and the cross hole are to be machined by using a turn-mill. This topic will be discussed at the end of the lesson.

Figure 9.3 Entities listed in the feature tree

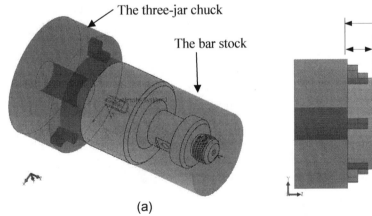

(a)

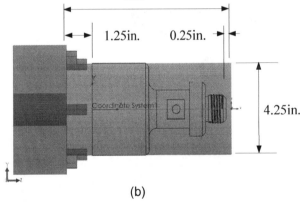

(b)

Figure 9.4 A bar stock of ϕ4.25in.×8in., (a) stock fixed in a chuck, and (b) stock size

The toolpath of the turn operations for machining the face, OD, and groove features usually consists of rough and finish operations. Hole drilling consists of center drill and drill operations. The rear end hole requires additional operations, such as bore, mainly due to its size. There is a total of fifteen operations created to machine this stub shaft. Toolpaths of the fifteen operations are shown in Figure 9.6. You may open the example file, *Stub shaft with toolpath.SLDPRT*, to preview the toolpath and turn operations created for this example.

9.3   Using CAMWorks

Open SOLIDWORKS Part

Open the stub shaft part (filename: *Stub shaft.SLDPRT*) downloaded from the publisher's website. This solid model shown in Figure 9.2 consists of nine solid features (revolve, chamfer, fillet, 3 cut extrudes, 2 cut revolves, and a thread), a point, two planes, and a coordinate system (see the solid feature tree in Figure 9.3). As soon as you open the model, you may want to check the unit system and increase the decimals from the default 2 to 4 digits.

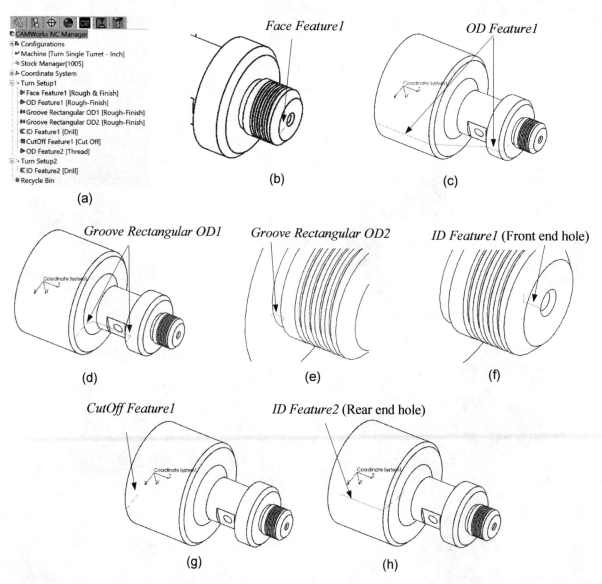

Figure 9.5 The machinable features of the turning operations: (a) CAMWorks feature tree, (b) *Face Feature1*, (c) *OD Feature1*, (d) *Groove Rectangular OD1*, (e) *Groove Rectangular OD2*, (f) *ID Feature1* (front end hole), (g) *CutOff Feature1*, and (h) *ID Feature2* (rear end hole) of *Turn Setup2*

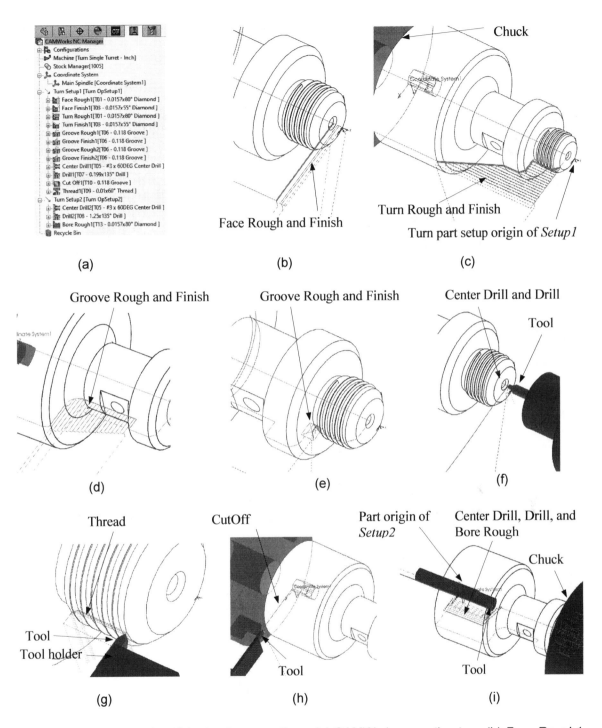

Figure 9.6 The toolpaths of the turning operations, (a) CAMWorks operation tree, (b) *Face Rough1* and *Face Finish1*, (c) *Turn Rough1* and *Turn Finish1*, (d) *Groove Rough1* and *Groove Finish1*, (e) *Groove Rough2* and *Groove Finish2*, (f) *Center Drill1* and *Drill1*, (g) *Thread1*, (h) *Cut Off1*, and (i) *Center Drill2*, *Drill2*, and *Bore Rough1* of *Turn Setup2*

Select NC Machine

Click the CAMWorks feature tree tab [CW] . Right click *Machine* and select *Edit Definition*. Similar to those of Lesson 8, under *Machine* tab of the *Machine* dialog box, we select *Turn Single Turret* from the list of *Applicable machines* box, and click *Select*. We choose *Tool Crib2 Rear* under *Available tool cribs* of the *Tool Crib* tab, select *T2AXIS-TUTORIAL* under the *Post Processor* tab, and select *Coordinate System1* for *Main spindle coordinate system* under the *Setup* tab.

Create Stock

From CAMWorks feature tree [CW], right click *Stock Manager* and choose *Edit Definition*. In the *Stock Manager* dialog box (Figure 9.7), we enter 8in. for the stock length, increasing by 0.25in. to the right of the part from that of Lesson 8. We enter the length of the stock outside the part to be –1.25in., as shown in Figure 9.7. As a result, a 0.25in. of extra material appears to the right end face of the part. The default stock material is *1005*. Accept the stock by clicking the checkmark ✓ at the top left corner. The bar stock should appear in the graphics area similar to that of Figure 9.4.

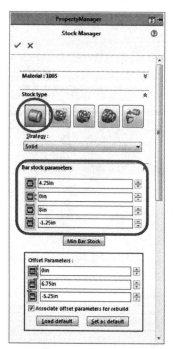

Figure 9.7 The *Stock Manager* dialog box

Turn Setups and Machinable Features

Click the *Extract Machinable Features* button [Extract Machinable Features] above the graphics area. Two setups, *Turn Setup1* and *Turn Setup2*, are created with multiple machinable features extracted. Under *Setup1*, there are six features extracted: *Face Feature1*, *OD Feature1*, *Groove Rectangular OD1*, *Groove Rectangular OD2*, *ID Feature1* and *CutOff Feature1*. Under *Setup2*, there is one feature extracted, *ID Feature2*. All machinable features are in magenta color, as shown in Figure 9.8. If you click any of these features, you should see them in the graphics area (as lines or curves) like those shown in Figure 9.5. Note that the *Thread1* solid feature was not extracted as a machinable feature. We will add a thread machinable feature manually later.

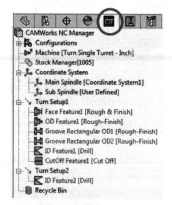

Figure 9.8 The machinable features extracted

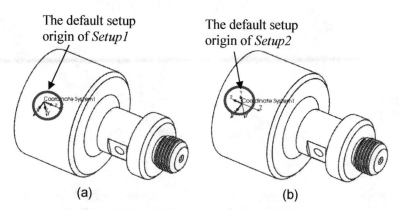

Figure 9.9 The default part setup origins of *Setup1* and *Setup2* at the origin of *Coordinate System1*

Click *Turn Setup1* and *Turn Setup2* to locate the respective setup origins in the graphics area. As shown in Figure 9.9, both origins are located at the origin of *Coordinate System1* with Z-axis points in opposite directions. The directions are all good. We will move the origin of *Setup1* to the front end face of the stock. Before we make this change, we will generate operation plans first.

Generate Operation Plan and Toolpath

Click the *Generate Operation Plan* button ⬛ above the graphics area.

Eleven and two operations are generated for *Turn Setup1* and *Turn Setup2*, respectively (Figure 9.10). All are in magenta color. Expand the operation, for example, *Turn Rough1*, to see the associated feature of the operation (in this case, *OD Feature1*). Click *OD Feature1* to show the feature in the graphics area, as shown in Figure 9.10.

Click the *Generate Toolpath* button 🔧 above the graphics area to generate toolpath. Turning toolpaths are generated for all operations.

Relocate Part Setup Origin for *Setup1*

As mentioned earlier, we will move the origin of *Setup1* to the center point of the front end face of the stock, which is more practical since this point is easy to access in setting up the stock on a lathe.

Under the CAMWorks operation tree tab ⬛, right click *Turn Setup1* and choose *Edit Definition*. In the *Operation Setup Parameters* dialog box, choose *Origin* tab, and select *Stock Vertex*. The vertex at the right end of the stock (circled in Figure 9.11) is selected. The part setup origin moved to the center of the front end face of the stock—see Figure 9.12(a)—as desired. We keep the part setup origin of *Setup2* as it is, as circled in Figure 9.12(b).

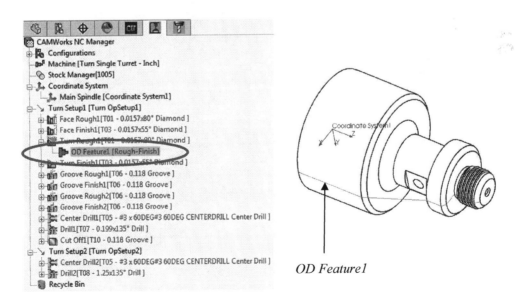

Figure 9.10 The thirteen operation plans generated

Click *OK* in the *Operation Setup Parameters* dialog box to accept the change. Click *Yes* to the question in the warning box: *The origin or chuck/fixture location/avoidance parameters has changed, toolpaths need to be recalculated. Regenerate toolpaths now?*

Review the Operations

Next we review the tool and key machining parameters of a selected operation, *Groove Rough1*.

Under the CAMWorks operation tree tab ![icon], right click *Groove Rough1* and choose *Edit Definition*. In the *Operation Parameters* dialog box, choose *Tool* tab and then *Groove Insert* tab, a groove insert of 0.118in. in width (0.118 Wide GROOVE INSERT) appears (Figure 9.13). Tool holder with groove insert appears in the graphics area (Figure 9.14). Click the *Holder* tab (see Figure 9.15); a standard holder of shank width 0.75in. and length 4in. has been chosen. Also, the *Down left* is chosen for *Orientation*, which defines the orientation of the cutter suitable to turn the groove feature.

Choose the *Groove Rough* tab of the *Operation Parameters* dialog box. In the *Parameters* area, the *Stepover* is set to *0.05in*, and *Allowance* is set to *0.01in* for both X and Z directions, as shown in Figure 9.16.

The *Stepover* defines the distance of the tool movement along the X-direction, similar to the depth of cut in milling operation. Allowances leave a small amount of material uncut, similar to those of rough mill operations. Click *Cancel* to close the dialog box since we are not making any changes to the *Groove Rough1* operation. You may select other operations to acquire a better understanding of the tool and parameters selected.

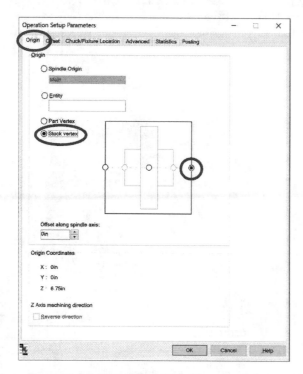

Figure 9.11 The *Origin* tab of the *Tool* tab in the *Operation Setup Parameters* dialog box

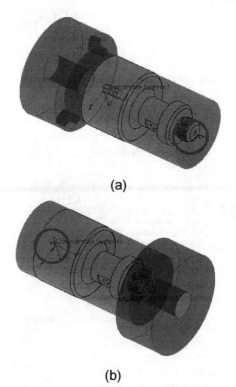

(a)

(b)

Figure 9.12 The part setup origins of (a) *Turn Setup1*, and (2) *Turn Setup2*

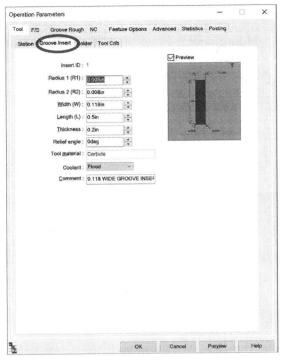

Figure 9.13 The *Groove Insert* tab of the *Tool* tab
of the *Operation Parameters* dialog box

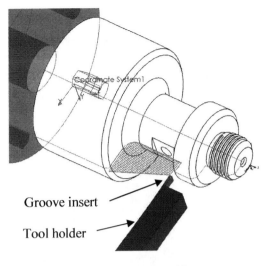

Groove insert

Tool holder

Figure 9.14 Tool holder with groove
insert appearing in the graphics area

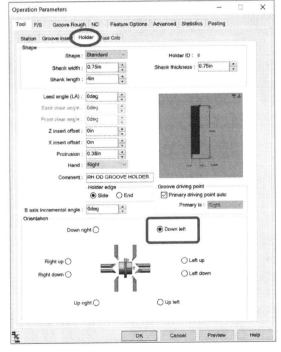

Figure 9.15 The *Holder* tab of the *Tool* tab of
the *Operation Parameters* dialog box

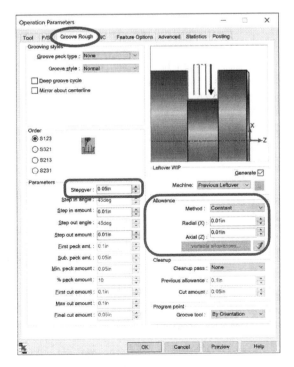

Figure 9.16 The *Groove Rough* tab of the
Operation Parameters dialog box

Simulate Toolpath

Click the *Simulate Toolpath* button 🔧 above the graphics area to simulate toolpath. The material removal simulation appears similar to those of Figure 9.17(a) and Figure 9.17(b) for *Turn Setup1* and *Turn Setup2*, respectively. We will add a thread operation to *Setup1*, and add a bore operation to *Setup2* for enlarging the hole at the rear end.

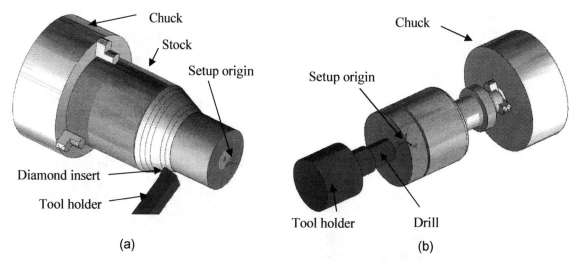

(a)   (b)

Figure 9.17 The material removal simulation, (a) *Turn Setup1*, and (2) *Turn Setup2*

9.4   Cutting the Thread

We now take a look at the thread. Since the thread solid feature was not extracted as a machinable feature, we will have to manually create one.

From CAMWorks feature tree 🔳, right click *Turn Setup1* and choose *Turn Feature* (Figure 9.18).

In the *New Turn Feature* dialog box (Figure 9.19), choose *OD Feature* for *Type* (should have been chosen by default), select *Thread* for *Strategy*, and then pick the line segments that represent the crest of the thread near the front end of the part in the graphics area (see Figure 9.19). The line segments picked are now listed under *Selected Entities* of the *New Turn Feature* dialog box. And a thread profile is created automatically by joining these line segments (joints are shown in the part and listed in the *Selected Entities*). Click the checkmark ✔ to accept the thread feature.

In the CAMWorks feature tree 🔳 , an *OD Feature2 [Thread]* is added and is in magenta color under *Turn Setup1*, as shown in Figure 9.20.

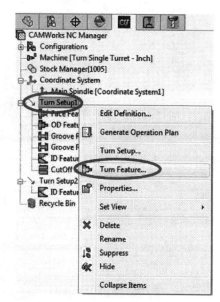

Figure 9.18 Creating a threading machinable feature

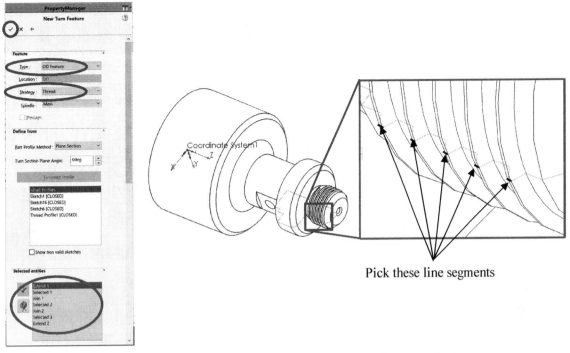

Figure 9.19 Picking the line at the front end for a turn feature

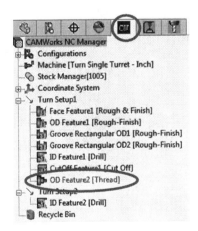

Figure 9.20

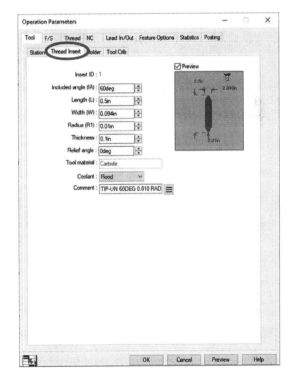

Figure 9.21 The *Thread Insert* tab of the *Tool* tab
of the *Operation Parameters* dialog box

Right click *OD Feature2 [Thread]* and choose *Generate Operation Plan*. A *Thread1* operation is now added to the operation tree under *Turn Setup1*, again in magenta color.

Right click *Thread1* and choose *Generate Toolpath*. Toolpath will be generated similar to that of Figure 9.6(g).

Next we take a closer look at the thread operation and thereafter, the toolpath.

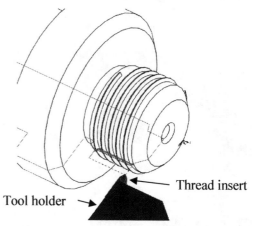

Under the CAMWorks operation tree tab [icon], right click *Thread1* and choose *Edit Definition*. In the *Operation Parameters* dialog box, choose *Tool* tab and then *Thread Insert* tab; a thread insert of $0.01 \times 60°$ (radius 0.01in. included angle 60 degrees, thickness 0.1in.) appears (Figure 9.21). Tool holder with thread insert appears in the graphics area (see Figure 9.22). Click the *Holder* tab (see Figure 9.23); a standard holder of shank width 0.75in. and length 4in. has been chosen. Also, the *Down left* is chosen for *Orientation*, which defines the orientation of the cutter suitable to turn the thread feature.

Figure 9.22 Tool holder with thread insert appearing in the graphics area

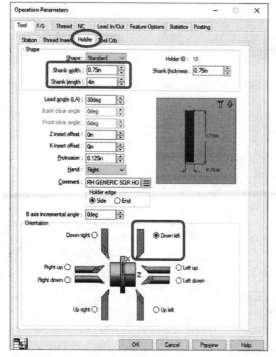

Figure 9.23 The *Holder* tab of the *Tool* tab of the *Operation Parameters* dialog box

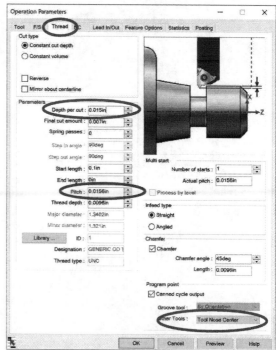

Figure 9.24 The *Thread* tab of the *Operation Parameters* dialog box

Click the *Thread* tab (Figure 9.24), and note that the *Depth per cut* is *0.015in.*, *Pitch* is *0.0156in.* (which is inconsistent with that of the solid feature and will be corrected next), and *Tool Nose Center* is chosen for *Program point*. The other option for *Program point* is *Tool Nose Tip*. Different options affect the G-code.

Click the *Feature Options* tab and click *Parameters* (see Figure 9.25). In the *OD Profile Parameters* dialog box (Figure 9.26), enter *0.0451in.* for *Thread depth* and *0.125in.* for *Pitch* so as to make the thread size consistent with that of the solid feature—see dimensions of the thread in Figure 9.2(c).

Note that *Maximum dia (D1): 1.3420in* and *Minimum dia. (D2): 1.25in* circled in Figure 9.26 are the major and minor diameters of the thread, respectively, extracted by CAMWorks.

Click *OK* in the *OD Profile Parameters* dialog box. The *Pitch* is now *0.125in.* under the *Thread* tab of the *Operation Parameters* dialog box. Click *OK* in the *Operation Parameters* dialog box to accept the changes. Toolpath will be regenerated.

Step Through Toolpath

Now we take a closer look at the toolpath to better understand the toolpath (and later the G-code) of the threading operation.

Right click *Thread1* and choose *Step Thru Toolpath*. The *Step Through Toolpath* dialog box appears (Figure 9.27).

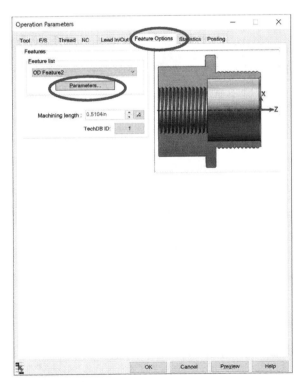

Figure 9.25 The *Feature Options* of the
Operation Parameters dialog box

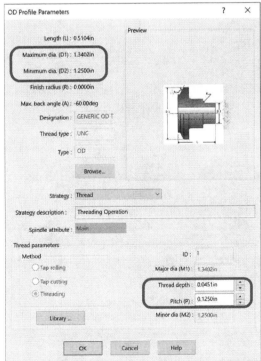

Figure 9.26 The *OD Profile Parameters*
dialog box

Under *Information*, CAMWorks shows the tool movement from the current to the next steps, the feedrate, and spindle speed, among others.

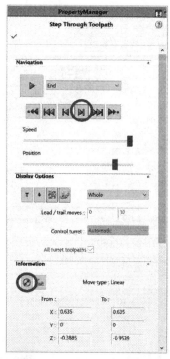

Click the *Step* button ◉ at the center (circled in Figure 9.27) to step through the toolpath. You may want to turn on *Show toolpath points* and *Tool Holder Shaded Display* to see the toolpath display similar to that of Figure 9.28.

A zoom in view of the toolpath (see Figure 9.28) indicates that the toolpath follows the trace of the tool center (center of the tool nose radius of the insert). As a result, in general, the toolpath offsets an amount of tool radius from the part boundary. For example, the XZ coordinates of the last cutting pass, as shown in Figure 9.29(a), are from (0.635, −0.3885) to (0.635, −0.9539). You may click the *Radial or Diameter Coordinate Display* button ◎ (circled in Figure 9.27) to toggle the numeric values displayed for the X-coordinate between radius and diameter.

The X coordinates indicate that the last pass is 0.635in. away from the part set up origin, as shown in Figure 9.29(a), which is the minor radius of the thread plus the tool radius; i.e., 0.625+0.01 = 0.635, as it should be.

Recall that we chose *Tool Nose Center* for *Program point*; hence, XYZ coordinates of the toolpath are generated at the tool nose center. If you choose *Tool Nose Tip* for *Program point*, the X coordinate of the last cutting pass will be 0.625, which is simply the minor radius of the thread.

Figure 9.27 The *Step Through Toolpath* dialog box

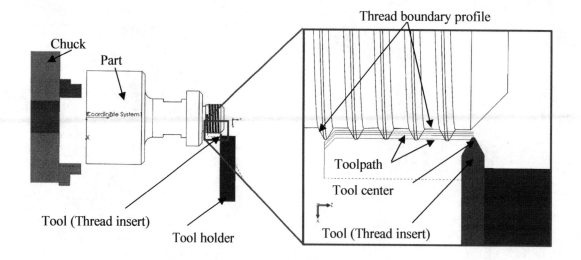

Figure 9.28 Toolpath of *Thread1*, a closer look

G-Code of the Threading Operation

Right click *Thread1* and choose *Post Process* to generate the G-code. Enter *Stub shaft thread* for filename.

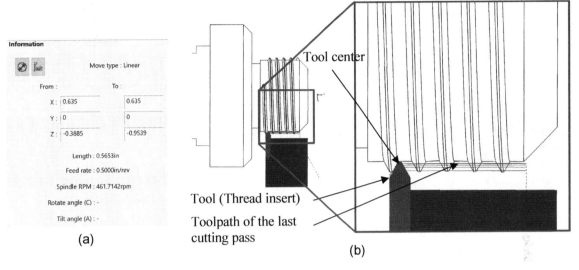

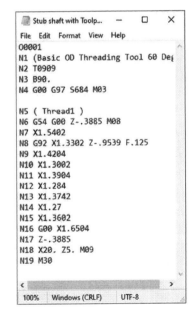

Figure 9.29 The last cutting pass of the toolpath of *Thread1*, (a) the XYZ coordinates of the toolpath, and (b) a zoom-in view of the cutter location

Figure 9.30 The G-code of *Thread1* operation

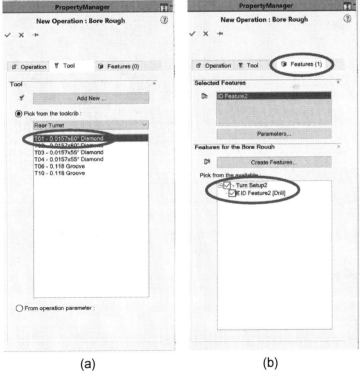

Figure 9.31 The *New Operation: Bore Rough* dialog box, (a) the *Tool* tab, and (b) the *Features* tab

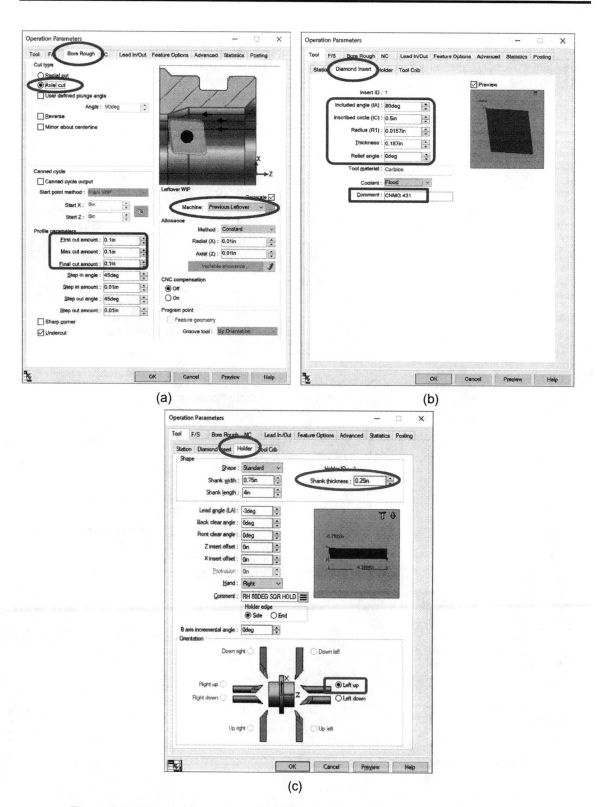

(a)

(b)

(c)

Figure 9.32 The *Operation Parameters* dialog box, (a) the *Bore Rough* tab, (b) the
Diamond Insert tab, and (c) the *Holder* tab

Open *Stub shaft thread.txt* file (see the file contents shown in Figure 9.30). It is shown that the NC block N14 moves the cutter along the last cutting pass, in which the X-coordinate is 1.27 (= 2×0.635), which is correctly converted.

9.5   Boring the Hole

We now add a boring operation to the hole at the rear end under *Turn Setup2*. Under the CAMWorks operation tree tab , right click *Turn Setup2* and choose *Turn Bore Operations* and then *Bore Rough*. In the *New Operation: Bore Rough* dialog box, choose *T01 – 0.0157X80° Diamond* for tool [see Figure 9.31(a)], and click the *Features* tab to choose *ID Feature2 [Drill]* as shown in Figure 9.31(b). Click the checkmark ✔ to accept the operation.

Click *OK* to the CAMWorks message: *The current insert/holder orientation is not allowed for the Main Spindle while using the Rear 1 Turret. CAMWorks has selected a valid orientation. Please review this selection on the Holder page and change it if required.* This is because the same tool (T01) was used for operations of *Turn Setup1*, in which *Down left* has been chosen for orientation. The down left orientation is not suitable for the hole boring operation and must be changed.

In the *Operation Parameters* dialog with the *Bore Rough* tab selected, we stay with the default values and selections (for example, *Axial cut, 0.1in.* for *First, Max,* and *Final cut amount*). Also, we choose *Previous Leftover* for *Machine*—see Figure 9.32(a)—which is left over from the *Drill2* operation.

Click the *Tool* tab, and then *Diamond Insert* tab; see Figure 9.32(b). We use the insert currently selected (*CNMG 431*).

Click the *Holder* tab; see Figure 9.32(c). We reduce the shank thickness to *0.25in.* to avoid collision. Note that *Left up* is chosen for *Orientation*, which is what the earlier message was referring to.

Click *OK* to accept the operation and click *Add* to the warning message: *Tool parameters have been changed. Select Add to create a new tool. Select Change to modify the tool for all operations sharing this tool.* A new operation, *Bore Rough1*, is now listed in the operation tree.

Under the CAMWorks operation tree tab , right click *Bore Rough1* and choose *Generate Toolpath*. A bore rough tool path is shown in the graphics area, similar to that of Figure 9.6(i).

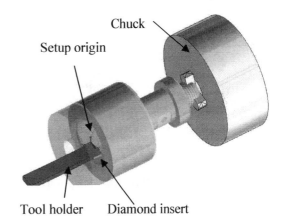

Figure 9.33 The material removal simulation of *Turn Setup2*

Choose *Simulate Toolpath* to carry out a material removal simulation, especially for *Turn Setup2* like that of Figure 9.33.

We have completed the first part of the exercise. You may want to save your model under a different name for future reference. Close the model.

Next, we will start over using a turn-mill machine that is able to cut both turn and mill machinable features.

9.6   Using a Turn-Mill

We will start from the very beginning by opening the part file *Stub Shaft.SLDPRT*.

Select NC Machine

Click the CAMWorks feature tree tab ⬚ . Right click *Machine* and select *Edit Definition*. In the *Machine* dialog box (Figure 9.34), choose *Mill-Turn Single Turret - Inch* from the list of *Applicable machines* box, and click *Select*.

Choose *the Tool Crib* tab, select *MT Tool Crib 2 Rear*, and then click *Select*. Choose the *Post Processor* tab; a post processor called *MT2AXIS-TUTORIAL* is selected. This is a generic post processor of the 2-axis mill-turn machine that comes with CAMWorks. Then, click the *Setup* tab and click *User Defined* under *Main spindle* and choose *Coordinate System1* for *Coordinate system*. Click *OK* to accept the selections and close the dialog box.

Create an identical bar stock; i.e., with outer diameter (4.25in.), inner diameter (0in.), overall length (8in.), length of the stock outside the part (−1.25in.), and stock material *1005*.

Extract Machinable Features

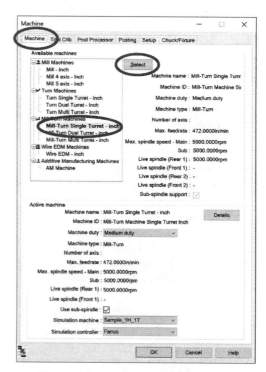

Figure 9.34 The *Machine* tab of the *Machine* dialog box

Click the *Extract Machinable Features* button ⬚ above the graphics area. In addition to *Turn Setup1* and *Turn Setup2*, two more setups are generated: *Mill Part Setup1* and *Mill Part Setup3*.

Under *Mill Part Setup1*, there is one feature extracted, *Hole1*; and Under *Mill Part Setup3*, there are two features, *Irregular Slot1* and *Irregular Slot2*, as shown in Figure 9.35. Machinable features extracted are also listed under *Turn Setup1* and *Turn Setup2*. We will focus on the mill operations and leave turn operations as they are since turn operations have been discussed in this lesson.

We first check the tool axes of the respective two mill part setups. Click *Mill Part Setup1* under the feature tree. A tool axis symbol ⬚ appears at the origin of the fixture coordinate system, *Coordinate System1*, with the arrow pointing in the X-direction of the fixture coordinate system, as shown in Figure 9.36(a), indicating the tool axis of the mill part setup. Since the cross hole is the only machinable feature extracted under *Mill Part Setup1*, the tool axis of the setup is adequate. No change is necessary.

Next, we click *Mill Part Setup3*. A tool axis symbol ⬚ appears again at the origin of the fixture coordinate system, but with the arrow pointing in the negative Z-direction of the fixture coordinate system; see Figure 9.36(b). The two machinable features under this setup are the two side cuts that must be machined by tools with the axis pointing in the X-direction. It is apparent that the tool axis of *Mill Part Setup3* is incorrect and must be changed.

Also, the two side cuts are located on the opposite sides of the shaft. They will have to be machined by operations under the two mill part setups, respectively.

We first attempt to modify the tool axis of *Mill Part Setup3* by right clicking it and choosing *Edit Definition*. In the *Mill Setup* dialog box (Figure 9.37), a face is selected, and in the graphics area, a tool axis symbol appears like that of Figure 9.36(b).

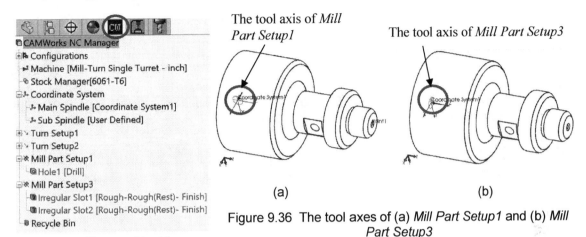

Figure 9.35 The machinable features extracted

(a)   (b)

Figure 9.36 The tool axes of (a) *Mill Part Setup1* and (b) *Mill Part Setup3*

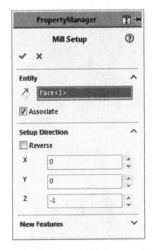

Figure 9.37 The *Mill Setup* dialog box

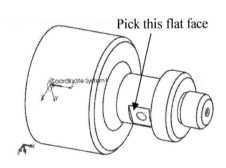

Figure 9.38 Picking the face of the side cut at the front side (*Irregular Slot1*)

Pick the flat face of the side cut at the front side (*Irregular Slot1*), as shown in Figure 9.38. A CAMWorks message appears: *This setup has 2 features. Cannot change the setup direction.* It indicates that changing the tool axis for *Mill Part Setup3* is not possible in CAMWorks. Click *OK* to close the message box. Close the *Mill Setup* dialog box.

We will create a mill part setup manually by right clicking *Stock Manager* and choosing *Mill Part Setup* (under CAMWorks feature tree ![icon]). The same *Mill Setup* dialog box appears. We now pick the flat face of the side cut at the front side, as shown in Figure 9.38. A tool axis symbol appears ⊁ at the fixture coordinate system, pointing in the negative X-direction (see Figure 9.39), which is adequate. Click the checkmark ✔ to accept the setup. A *Mill Part Setup4* appears in the feature tree.

You may extract machinable features for *Mill Part Setup4* by right clicking it and choosing *Recognize Features*. A machinable feature *Rectangular Slot1* is extracted and the side cut at the front is highlighted in the part (like that of Figure 9.40), indicating that the desired feature is extracted.

Right click *Rectangular Slot1* and choose *Generate Operation Plan* and then *Generate Toolpath*. Two operations, *Rough Mill1* and *Contour Mill1*, are generated, as shown in Figure 9.41(a) and Figure 9.41(b), respectively. Note that if you see a *Rough Mill2* operation created, you may change the strategy of the machinable feature *Rectangular Slot1* from *Rough-Rough(Rest)-Finish* to *Rough-Finish*. You may right click *Rectangular Slot1* (under CAMWorks feature tree 🗔) and choose *Edit Definition*. In the *2.5 Axis Feature* dialog box, click *Edit Feature*, choose *End Condition* tab, and select *Rough-Finish* for *Strategy*. Regenerate the operation plan, and then regenerate the toolpath.

You may show material removal simulation by first suppressing *Mill Part Setup4* and click the *Simulate Toolpath* button ⌖ Simulate Toolpath above the graphics area. The simulation should appear like that of Figure 9.41(c).

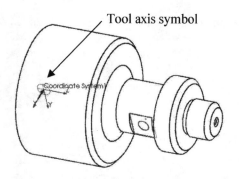

Figure 9.39 Tool axis symbol appearing at the fixture coordinate system, pointing in the negative X-direction

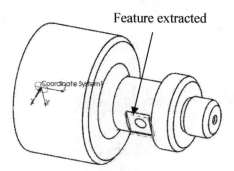

Figure 9.40 The machinable feature extracted

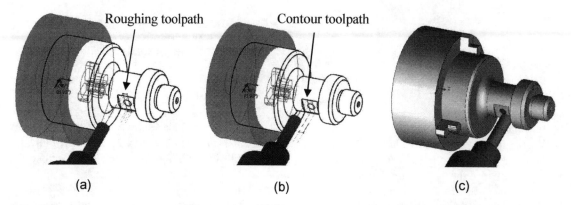

(a)   (b)   (c)

Figure 9.41 The toolpath of the operations under *Mill Part Setup4*, (a) *Rough Mill1*, (b) *Contour Mill1*, and (c) material removal simulation

Now we do the same to extract the side cut feature on the rear side by right clicking *Mill Part Setup1* and choosing *Recognize Features*. The feature at the rear side is extracted as *Rectangular Slot2*. As a result, there are two machinable features listed under *Mill Part Setup1*, including *Hole1* extracted earlier.

Generate operation plans and toolpaths for both machinable features. Four operations are created: *Center Drill1* and *Drill1* for *Hole1*, and *Rough Mill2* and *Contour Mill2* for *Rectangular Slot2*. Again, you may change the strategy for *Rectangular Slot2* from *Rough-Rough(Rest)-Finish* to *Rough-Finish*.

You may create a material removal simulation to simulate the entire machining operation, including turning and milling, as shown in Figure 9.42. Note that the chuck is slightly misplaced in the material removal simulation of *Mill Part Setup1* and *Mill Part Setup4*. You may delete *Mill Part Setup3*.

We have now completed the exercise. You may save your model for future reference.

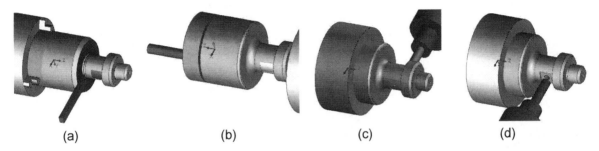

(a)   (b)   (c)   (d)

Figure 9.42 The material removal simulation, (a) *Turn Setup1*, (b) *Turn Setup2*, (c) *Mill Part Setup1*, and (d) *Mill Part Setup4*

9.7   Exercises

Problem 9.1. Create NC operations to machine the part shown in Figure 9.43 from a stock of ϕ2.75in.×4in. Pick adequate tools (with justifications). Please submit the following for grading:

(a)   A summary of the turn operations, including cutting parameters and tools selected.
(b)   A summary of the mill operations, including cutting parameters and tools selected.
(c)   Screen shots of combined NC toolpaths and material removal simulations.
(d)   Is there any material remaining uncut? If so, was it a part design issue or a machining issue? What can be done to improve either the part design or machining operations to correct the issue?

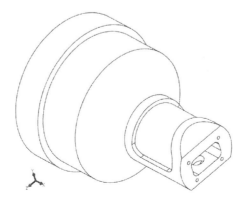

Figure 9.43 The design model of Problem 9.1

[Notes]

Lesson 10: Die Machining Application

10.1 Overview of the Lesson

In this lesson, we present an industrial application that involves die machining of tooling manufacturing for sheet metal forming. In this application, manufacturing of tooling for sheet metal forming, including punch and die, was carried out by using a HAAS mill. CAMWorks was employed to conduct virtual machining and toolpath generation for machining the die and punch.

We start by briefly discussing the sheet metal part that is to be formed, as well as the tooling design and the sheet metal forming simulation. We then discuss the steps of adding a HAAS mill to CAMWorks in support of machining simulation and post processing for G-code. We only discuss the virtual machining of the die to avoid repetition. At the end of the lesson, we briefly discuss the implementation of the forming process using the manufactured tooling for forming the part at the shop floor. The goal of the lesson is to provide readers a flavor of the practical aspect of virtual machining carried out in CAMWorks for support of a practical application. In addition, we discuss the detailed steps in adding a virtual HAAS mill and an associated post processor to CAMWorks. You may acquire computer files of virtual machines, similar to those to be discussed in Section 10.5, from your machine vendor, and follow the same steps in Section 10.5 to add a virtual machine to CAMWorks that replicates the physical machine on your shop floor. The HAAS machine files mentioned in the lesson are not provided to the readers due to the copyright issue.

The part files, including *part_1_draw_die_r1.SLDPRT* and *part_1_draw_die_r1 with toolpath.SLDPRT*, are available at the publisher's website for download. You may open the model file with toolpath; i.e., *part_1_draw_die_r1 with toolpath.SLDPRT*, to preview the toolpaths created for the die block. You may also open the model file, *part_1_draw_die_r1.SLDPRT*, and follow the discussion in Section 10.7 to create toolpaths and machining simulation similar to those discussed in Section 10.8. Instead of using a HAAS virtual machine, you may use your own virtual machine or the legacy machine, for example, *Mill_Tutorial*, that came with CAMWorks, to carry out machining simulation using the Machine Simulation capability of CAMWorks.

10.2 The Sheet Metal Part Design

The sheet metal part to be formed is a half clamp of a fuel line in an aerospace engineering system. The half clamp part shown in Figure 10.1(a) is symmetric in geometry, and is made of Alclad aluminum alloy 2024 with thickness 0.05 in. Key dimensions of the part are given in Figure 10.1(b). As can be expected, forming the part in one shot can be difficult due to the double curvature around the neck and the bends at the ears, which induce both tearing and wrinkles during the forming process. Usually, such a part with delicate geometry requires multiple steps (e.g., bending the ears after the main body is formed). However, since one-shot forming saves both man hours and tooling cost, the focus was to develop a forming process and tooling that form the part to its desired shape through one single forming operation.

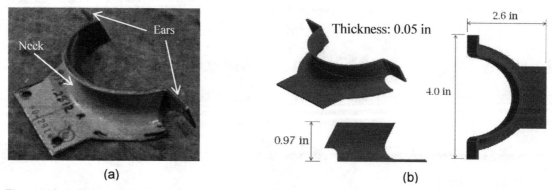

Figure 10.1 The half clamp part with double curvature (a) physical sample part, (b) CAD model with key dimensions.

10.3 Sheet Metal Forming Simulation

Forming simulations were first conducted to explore the formability of the part. The die face, including the surface of the two deep pockets in the die, was designed based on the geometric shape of the sheet metal part.

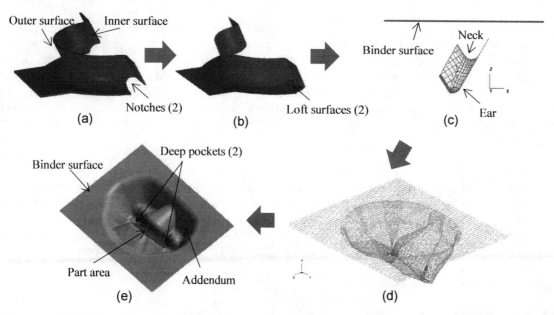

Figure 10.2 Die face design, (a) solid model of the sheet metal part, (b) mid-surface extracted and notches filled with loft surfaces, (c) flat binder surface created (side view), (d) addendum created, and (e) the complete die face.

The mid-surface of the part was first extracted as a surface model in SOLIDWORKS by compressing the outer and inner surfaces of the solid model; see Figure 10.2(a). The notches near the ears of the surface model were filled with loft surfaces; see Figure 10.2(b). The revised surface model was brought into DynaForm (www.dynaform.com) for die face design in support of forming simulations. The part was carefully oriented to prevent a potential undercut. In the meantime, a flat binder surface was created above the part; see Figure 10.2(c). An addendum surface that connects the binder surface and the part area was then automatically generated in DynaForm; see Figure 10.2(d). By trimming the binder surface with the addendum, the complete die face, shown in Figure 10.2(e), consisting of the part area, the addendum, and the trimmed binder surface, was obtained.

The die face was used as the female tool, while the male punch was modeled by copying and offsetting the geometry of the die face. Figure 10.3 illustrates a setup of draw forming simulation, in which the blank was pressed down by binders. Note that the geometric shape of the blank shown in Figure 10.3 was obtained after numerous unsuccessful blank designs, in which only half of the blank was modeled due to symmetry.

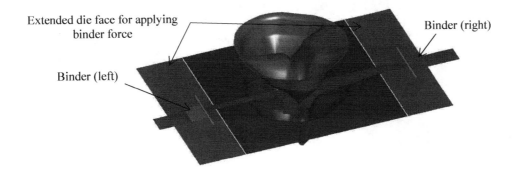

Figure 10.3 Draw forming simulation setup with binders

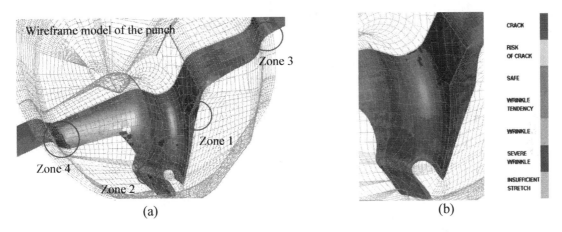

Figure 10.4 Draw forming simulations (with part boundary sketched in orange color in the half blank), (a) four zones with tearing (red color), and (b) successful forming simulations with modified die surface and blank shape (only mild wrinkles around the neck in pink color)

Initial forming simulations like that of Figure 10.4(a) reveal four areas with severe tearing. Tearing at Zones 3 and 4 is clearly due to the sharp transition in the die face geometry entering the cavity. The transition area of the die face was then modified to allow a smoother transition in geometry that facilitates material flow during forming. Tearing at Zones 1 and 2 is caused by the fact that the ears were bent early in the forming process before being dragged by the punch down to the deep pockets of the die, resulting in insufficient material flow from the right end, and hence excessively stretching the blank material to the left of the ears. The blank shape was further modified aiming at allowing the ear to slide around the tip of the punch without being bent, which would contribute to a more evenly distributed material flow. This turned out to be a key factor that led to a successful forming of the part at the shop floor. Forming simulation of the modified tooling and blank is shown in Figure 10.4(b), in which tearing is completely eliminated and yet only mild wrinkling appears around the neck area. Such simulations show sufficient promise that merits moving forward to the next stage in the development, that is, the tooling design.

10.4 Tooling Design

The finalized die face geometry was exported from DynaForm and imported into SOLIDWORKS for tooling design. The tooling was designed to support a forming scenario, in which the blank is fixed at its left end and pressed at its right end by binder. To implement the scenario, the blank is extended in the length direction, and its left end is wrapped over the side face of the die block, as illustrated in Figure 10.5(a) and (b). The left end of the blank was designed to be bent three times and held tightly against the die block with two clamps and twelve bolts in order to avoid slippage during forming. Two bolt holes were drilled at the right side of the top face of the die block, as shown in Figure 10.5(c), so that the clamp could be installed on the top face of the die block to provide a holding force that could be adjusted by tightening up the two bolts at different torque levels. In addition, two shallow slots were cut at the edges of the block, as shown in Figure 10.5(d), to facilitate the transverse alignment of the blank.

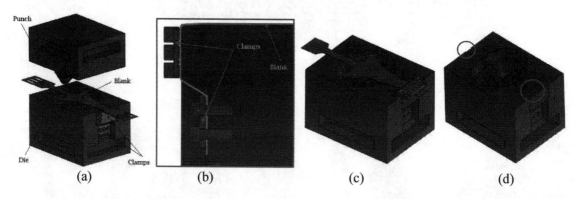

Figure 10.5 Tooling designed in SOLIDWORKS, (a) positioning of blank, punch, die, and clamps, (b) the left side of the blank is bent and held by two clamps to mimic a fixed boundary condition, (c) the right side of the blank is held by a clamp on the top face of the die block to mimic a binder condition, and (d) shallow slots designed for transverse blank alignment

10.5 Adding a Virtual CNC Mill to CAMWorks

We have access to HAAS mills at our machine shop at the university. Therefore, we first added a virtual HAAS mill to CAMWorks to facilitate virtual machining simulation for tooling manufacturing. Adding a virtual mill to CAMWorks supports the following. First, it allows the user to use a virtual machine in CAMWorks that is a replica of the physical machine at the shop floor. Second, it offers users a customized post processor that converts toolpath to G-code ready for machining without (or with minimum) modification. And third, it allows the virtual machine to be used in Machine Simulation like that of Lesson 7, in particular, detecting collision in machining operations.

Depending on the machine type there are at least three files needed for bringing in a customized mill for machine simulation. Files of extensions .ctl and .lng contain information on a customized post processor. The file with .kin extension defines the kinematics of the virtual machine. You will have to acquire these files for the CNC machine you are using from your machine vendor. These files are added to the folder

C:\CAMWorksData\CAMWorks2023X64\Posts.

If you list the contents of the folder, you should see several tutorial post processors came with CAMWorks listed, such as *M3AXIS-Tutorial*, *T2AXIS-Tutorial*, and *MT2AXIS-Tutorial*, that we used in the previous lessons.

We added three files to this folder: *Haas-5ax.ctl*, *Haas-5ax.kin*, and *Haas-5ax.lng*. You may do the same if you have acquired similar files from your machine vendors.

Note that in the folder, you may see files with extensions .pinf and .rtf. Files with .pinf extension are optional, which determines the default output file extension of the NC programs. The .rtf file is also optional, which displays information about the post processor when you select it in CAMWorks.

In addition, to support machine simulation, a set of geometric files that support the visualization of the virtual machine was added to the folder

C:\CAMWorksData\CAMWorks2023X64\MachSim\XML.

For this lesson, we added a folder *5AxHaasVF3-XT-210-trunnion* that contained files of the machine geometric information for the support of visualization of the virtual machine, in our case, HAAS 5-axis mill. You will have to acquire these files for the CNC machine you are using from your machine vendor. There are several subfolders under *XML* that came with CAMWorks, including *Mill_Tutorial* and *MillTurn_Tutorial*, which are generic machines supporting machine simulation. We chose one of them, *Mill_Tutorial*, for machine simulation in Lesson 7.

10.6 Customizing Technology Database

After adding files to CAMWorks data folders, we added the machine to CAMWorks by customizing its technology database, TechDB™. Here is what we did.

From the SOLIDWORKS pull-down menu, choose *Tools > CAMWorks > Technology Database*.

The *CAMWorks 2023 Technology Database* dialog box appears; see Figure 10.6(a). Click *Mill*, circled in Figure 10.6(a), choose *Inches*, and then select *Mill 5 axis – Inch*—the machine information appears in the box to the right; see Figure 10.6(b).

Click the *Copy* button—circled in Figure 10.6(b) at the top, to create a copy of the 5 axis mill. Enter the machine name, for example: *HAAS 5-axis Mill*, and a short machine description; see Figure 10.6(c). Then pull down the *Post processor* to choose *Haas-5ax.ctl* (which was added to the post processor folder at C:\CAMWorks data).

You may click other tabs, such as *Specifications*, to add more information for the new mill being added to CAMWorks. Click the *Save* button to add the HAAS mill.

You may modify tools to match with those at the shop floor by clicking, for example, *Mill Tooling* and choosing, for example, *Flat End Mill*—Figure 10.7(a)—to bring up the *Tool Database*, as shown in Figure 10.7(b).

You may choose a cutter and click the *Copy* button, as circled in Figure 10.7(b), and modify it to create a new flat-end cutter. Click the *Save* button to add the new flat-end cutter.

You may modify feedrates and other machining parameters to reflect your practice at the shop floor. You may click the *Feed/Speed* button from the *CAMWorks 2023 Technology Database* window, which is circled in Figure 10.6(a), and choose, for example, *Feed/Speed Editor*. A *Material Library* window appears, as shown in Figure 10.8(a), with a total of 31 materials listed in the left column. Click a material, for example *1005*, to show information of the selected material.

You may add a new material to the list by clicking the ✚ button circled in Figure 10.8(a).

Click the *Feeds and Speeds* tab circled in Figure 10.8(a) to bring out the *Micro Estimating – Feeds and Speeds* dialog box; see Figure 10.8(b). You may modify the data by directly editing the number in a data cell. You may continue editing the data in the database, save the changes by clicking the *Save* button, or click the *Close* button ("x" at the top right corner) to close the *Micro Estimating – Feeds and Speeds* dialog box. Click *Close* to close the *Material Library* dialog box. Click the *Close* button ("x" at the top right corner) to close the technology database.

10.7   Virtual Machining Simulation using CAMWorks

In this section, we discuss virtual machining of the die, in particular the machining operations that cut the cavity of the die [Figure 10.9(a)] from a 12in.×9in.×7.5in. Kirksite block, as shown in Figure 10.9(b). The material removal simulation is shown in Figure 10.9(c).

We used the HAAS-5ax mill added to CAMWorks and generated four operations: volume roughing, local milling, and two surface millings. Note that two surface milling operations are required to achieve desired surface finish of a maximum scallop 0.0003in. We also picked the post processor *Haas-5ax.ctl* to convert the toolpath to G-code compatible with the HAAS mill at the machine shop.

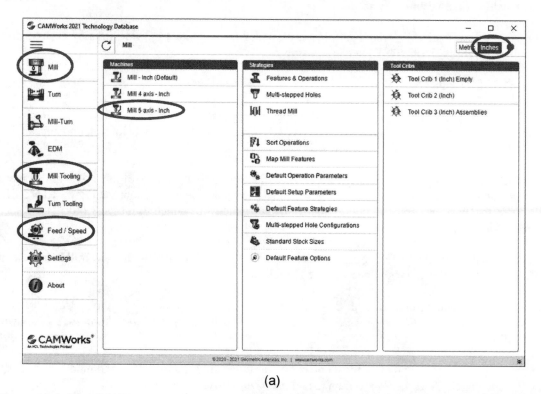

(a)

Figure 10.6 The *CAMWorks 2023 Technology Database* dialog box, (a) selecting *Mill* in the main window, (b) selecting *Mill 5 axis - Inch* and clicking *Copy*, and (c) entering machine information

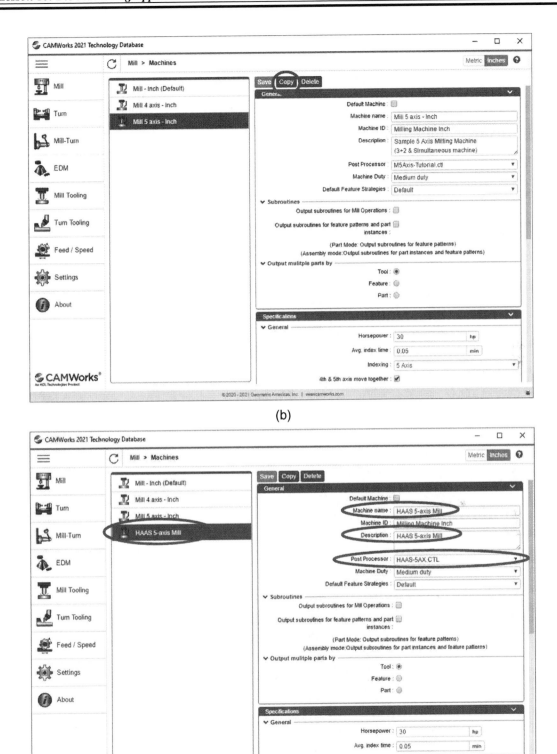

(b)

(c)

Figure 10.6 The *CAMWorks 2023 Technology Database* dialog box (cont'd)

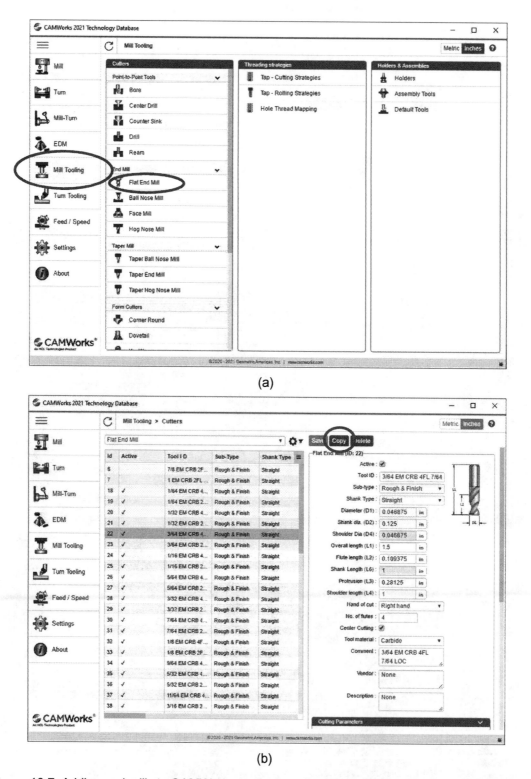

(a)

(b)

Figure 10.7 Adding end mills to CAMWorks technology database, (a) the *CAMWorks Technology Database* window, and (b) copying, editing, and saving a new end mill to the database

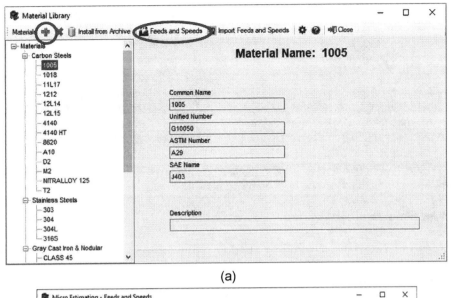

(a)

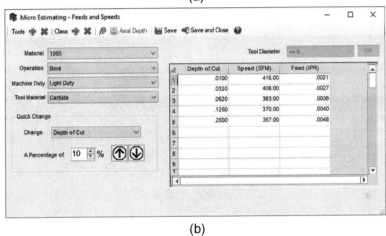

(b)

Figure 10.8 The *Micro Estimating Material Library* window, (a) materials listed under *Material List* tab, and (b) parameters under *Speeds & Feeds* tab

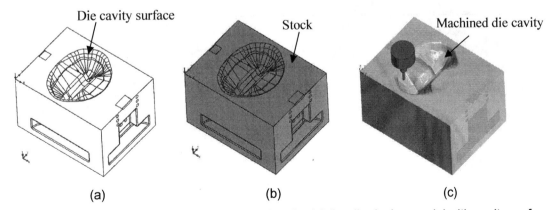

Figure 10.9 Virtual machining of the die block example, (a) the die design model with cavity surface, (b) the raw stock, a rectangular Kirksite block, and (c) the material removal simulation of the die cavity machining operations

You may open the model file downloaded from the publisher's website to browse the toolpath of the die machining. Again, the file name is *part_1_draw_die_r1 with toolpath.SLDPRT*.

The tools and key machining parameters employed for the four operations are listed in Table 10.1.

The following was what we did. We right clicked *Machine* and chose *Edit Definition*. In the *Machine* tab of the *Machine* dialog box, we selected *HAAS-5axis*. In the *Post Processor* tab, we chose *HAAS-5AX* for post processor. We then generated the four operations. The toolpath and material removal simulation of the four opertations are shown in Figure 10.10 and Figure 10.11, respectively.

Note that a 0.25in. ball-nose cutter is able to reach deep enough to the bottom of the two pockets for final clean up and surface polishing, as can be seen by comparing Figure 10.11(d) and Figure 10.11(e).

10.8   Machine Simulation

We chose the HAAS virtual machine added to CAMWorks to carry out the machine simulation.

Recall what we discussed in Lesson 7. To bring up the machine simulation, we right clicked *Mill Part Setup1* under CAMWorks operation tree ⬛, and chose *Machine Simulation* and then *Legacy*.

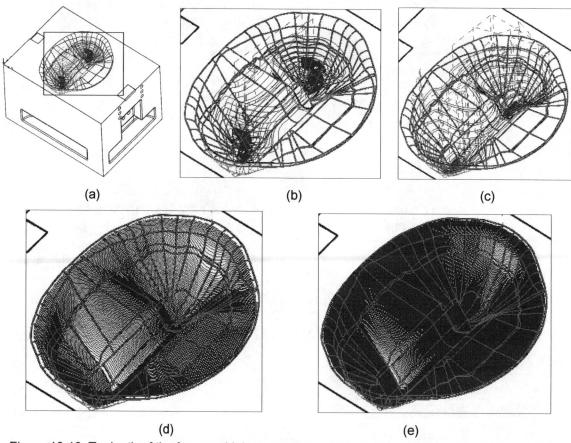

(a)   (b)   (c)

(d)   (e)

Figure 10.10 Toolpath of the four machining operations, (a) volume roughing, (b) volume roughing, a closer view, (c) local milling, (d) surface milling 1, and (e) surface milling 2

The default *Mill_Tutorial* machine appeared like that of Lesson 7. We selected the HAAS mill for simulation by pulling down the *Machine* selection and choosing *5AxHaasVF3-XT-210* as the machine (see Figure 10.12). Click the *Update* button on top (circled in Figure 10.13) to bring out the HAAS mill.

Then we clicked the *Select* button ▢ next to the *Machine* on top of the *Machine Simulation* window (circled in Figure 10.13) to bring up the *Select Point* dialog box (Figure 10.14).

We picked the coordinate system, *Coordinate System1*, in the graphics area (see Figure 10.15), and entered 6in., 4.5in., and -7.5in. for the X-, Y-, Z-offsets, respectively, in the *Select Point* dialog box (Figure 10.14).

Table 10.1 The tools and key machining parameters employed for the four NC operations

Operations	Tool	Stepover	Depth of cut	Scallop height
Volume Roughing	1 in. flat-end	0.25 in.	0.25 in.	
Local Milling	0.5 in. ball-nose	0.1 in.	0.1 in.	
Surface Milling 1	0.5 in. ball-nose			0.005 in.
Surface Milling 2	0.25 in. ball-nose			0.0003 in.

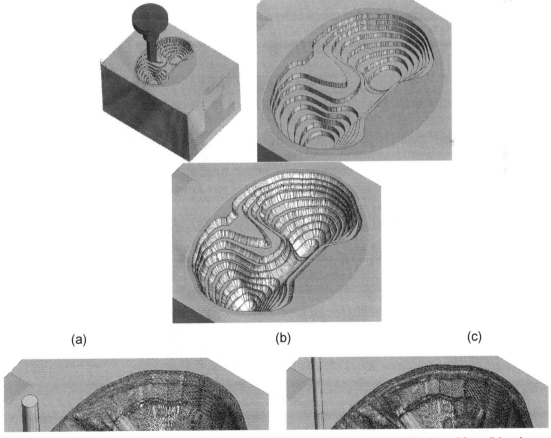

(a)   (b)   (c)

Figure 10.11 Material removal simulation of the four operations, (a) volume roughing, (b) volume roughing, a closer view, (c) local milling, (d) surface milling 1, and (e) surface milling 2

These selections positioned the stock at the center of the rotary table mounted on the tilt table in the HAAS mill like that of Figure 10.16.

We clicked the *Machine Housing* button 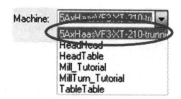 above the graphics area (circled in Figure 10.13) and chose *Hide* to hide the machine housing. We then rotated the view to see that the stock sat at the center of the top face of the rotary table, as shown in Figure 10.16.

Figure 10.12 Choosing *HAAS mill* as the machine

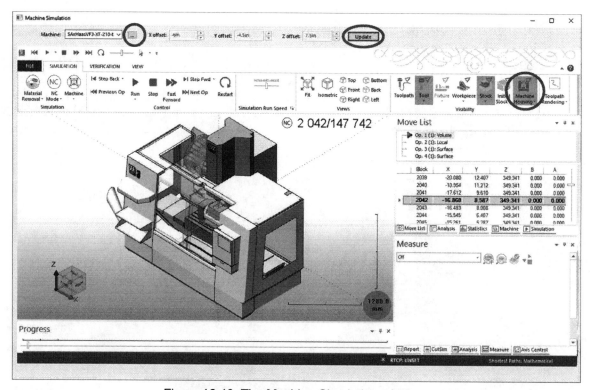

Figure 10.13 The Machine Simulation window

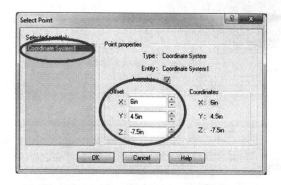

Figure 10.14 The *Select Point* dialog box

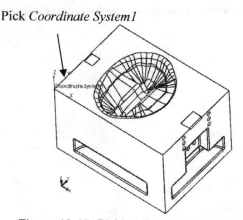

Pick *Coordinate System1*

Figure 10.15 Picking the coordinate system, *Coordinate System1*

We clicked the *Run* button ▶ to start the simulation. The NC blocks of the corresponding tool motion were displayed in the lower portion of the *Move List* area.

The simulation ran to the end for the first three operations without any issue (Figure 10.17). However, collisions were detected between the tool holder and the die surface when the 0.25in. ball-nose cutter was reaching the bottom of the pockets in the final operation, *Surface Milling 2*, as circled in Figure 10.18. To avoid the collisions, a longer cutter (4.5in. in length) and a tool holder of smaller diameter (0.75in.) were employed.

The same cutter and holder were used when we physically cut the die at the shop floor.

We followed the same steps as discussed in Lesson 2 to convert the toolpath into G-code by choosing the post processor *HAAS-5AX* (right click *Machine*, choose *Edit*, and click the *Post Process* tab).

10.9 Machining the Die Using HAAS Mill

In addition to the toolpath that cuts the die cavity as discussed above, other toolpaths were generated to machine the notches on the top face, slots on the four side faces, and tap the holes. Similar toolpaths were created to machine the punch as well.

The G-code converted was then uploaded to a HAAS mill to machine the die (and punch, see Figure 10.19). The machined punch and die (cut out from two Kirksite blocks) as well as one of the aluminum clamps are shown in Figure 10.20(a).

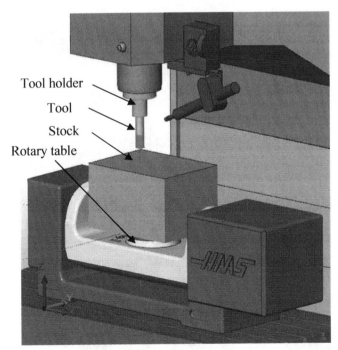

Figure 10.16 The stock sitting at the center of the top face of the rotary table

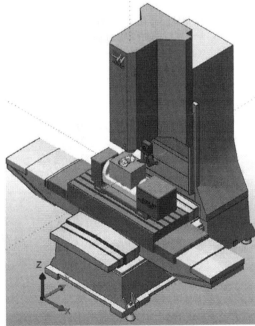

Figure 10.17 Running simulation for all four operations

The machined tooling blocks, including the die, punch, and clamps, are of excellent quality with desired surface finish. The tooling is ready for forming the part. Figure 10.20(b) shows the implementation of the scenario discussed in Section 10.3.

10.10 Process Implementation and Validation

Forming attempts were conducted on a 300 ton press at the shop floor. Tearing was found in all attempts near Zone 1 shown in Figure 10.4(a), indicating that the material flow was not sufficient. After the first failed attempts, a few more blanks were formed by adding a lubricant between the tooling and the blank to increase material flow, and at the same time lowering the blank holding force to minimum (bolts finger tightened). Although with the increasing material flow, smaller cracks still appeared on the blank at about the same location.

As a last resort, two forming tricks were implemented based on the finding in simulation that the ears on the blank cannot be bent and dragged into the deep pockets, which was believed to be the major cause of tearing on the failed blanks.

By bending the ears beforehand, the contact area between the ears and the punch was significantly reduced, which helped the ears to slide away from the tip of the punch, thus avoiding material being dragged into the deep pockets. In addition, to maximize the material flow from the right, the forming process was paused at about 0.75in. before closure between the punch and die. The partially formed blank was taken out of the tooling and trimmed off its right portion to completely remove holding force from the right. The trimmed blank was then brought back to the tooling and the forming was resumed until the punch and die closed completely. This process is illustrated in Figure 10.21.

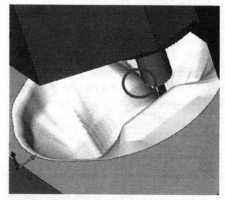

Figure 10.18 Collision encountered between the tool holder and the die surface

(a)

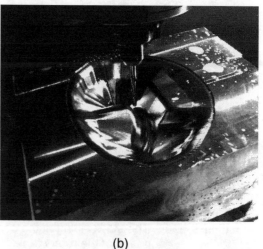

(b)

Figure 10.19 Machining the die on a HAAS mill, (a) the Kirksite block mounted on the jig table, and (b) a closer look at the machined die block

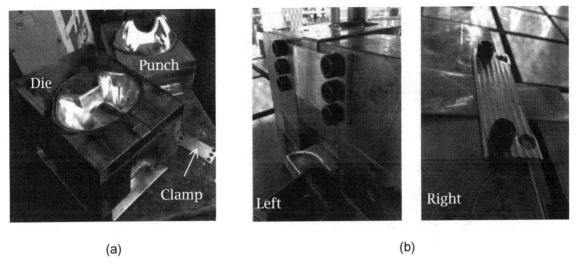

(a)   (b)

Figure 10.20 Manufactured tooling blocks, (a) punch, die and clamps machined, and (b) implementation of the forming scenario with blank fixed at its left side and clamped on its right side

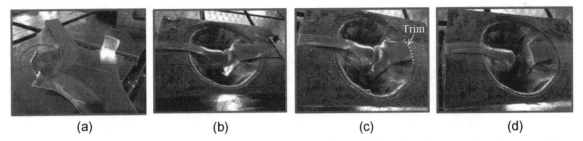

(a)   (b)   (c)   (d)

Figure 10.21 Forming tricks implemented at the shop floor: (a) bending the ears in advance, (b) forming the blank until the punch is 0.75 in. before closing, (c) trimming off the right side of the blank to allow more material flow, and (d) continuing forming until die and punch close.

Figure 10.22 Successfully formed part (left) with the matching half in yellow color (right).

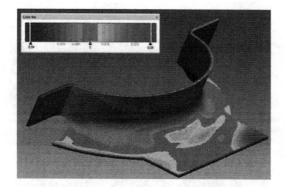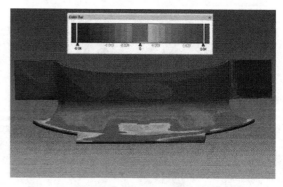

Figure 10.23 Accuracy verification for the formed part.

With the implementation of the two tricks, the ears were able to slide away without being dragged into the deep pockets. A blank was formed at the end with no tearing or wrinkle, producing a working part that fit well with the other half of the fuel line clamp, as shown in Figure 10.22. The same forming process was repeated several times and the same working parts were formed successfully, which verified the repetitiveness of the process.

The successfully formed part was scanned into the computer using ATOS III scanner (www.gom.com). The scanned geometry was compared with the CAD solid model of the part design for accuracy verification. As shown in Figure 10.23, the deviation between the original CAD design and the scanned part was within ±0.04 in, which satisfied the accuracy requirements for the part. The project came to a perfect closure at this point. CAMWorks has played a significant role in the success of the project.

Appendix A: Machinable Features

A.1   Overview

Feature-based machining is the driving concept behind CAMWorks. By defining areas to be machined as features, CAMWorks is able to apply more automation and intelligence into toolpath creation. In this appendix, we discuss machinable features that are extracted by using feature recognition capabilities provided by CAMWorks, including automated feature recognition (AFR) and interactive feature recognition (IFR) and local feature recognition (LFR). We focus on more machinable features for milling operations. Our discussion includes the NC operations generated for the machinable features by the rules, called strategy or machining strategy, implemented in CAMWorks technology database. We also briefly discuss machinable features for turning operations.

A.2   Machinable Features for Milling Operations

Toolpaths can be generated only on machinable features. CAMWorks provides three methods, AFR, IFR, and LFR, for defining machinable features.

Automatic Feature Recognition (AFR)

Automatic Feature Recognition analyzes the part geometric shape and attempts to extract most common machinable features such as pockets, holes, slots and bosses. You may select the *Extract Machinable Features* button ⬚ above the graphics area, choose from the pull-down menu *Tools > CAMWorks > Extract Machinable Features*, or right click *Mill Part Setup* (under CAMWorks feature tree ▣) and choose *Recognize Features* to initiate AFR. Depending on the complexity of the part, AFR can save considerable time in defining 2.5 axis features, including prismatic features commonly found in milling operations.

Most 2.5 axis features can be extracted automatically. One of the major characteristics of the 2.5 axis features is that the top and bottom of the feature are flat and normal to the tool axis of the machining operations, including prismatic solid features and solid features with tapered walls. Typical 2.5 axis features can be a boss, pocket, open pocket, corner slot, slot, hole, face feature, open profile, curve or engrave feature. Some of these features are illustrated in Figure A.1.

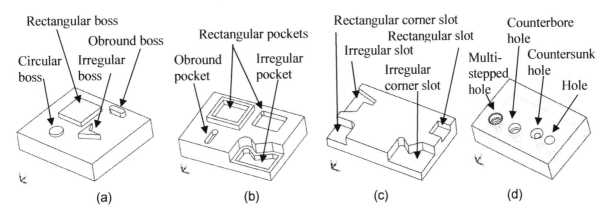

Figure A.1 Illustration of 2.5 axis features, (a) bosses, (b) pockets, (c) slots, and (d) holes

The types of 2.5 axis features to be included in AFR can be selected in CAMWorks. You may choose from the pull-down menu *Tools > CAMWorks > Options* to bring up the *Options* dialog box like that of Figure A.2.

If you click the *Mill Features* tab of the *Options* dialog box, the default feature types to be extracted are listed. By default, holes, non-holes (such as pocket, slot, etc.), boss, and tapered & filleted are selected.

Also, *MfgView* is selected for *Method*. Two methods are currently provided: *AFR* and *MfgView*.

When *AFR* is selected, CAMWorks analyzes the SOLIDWORKS solid model and identifies two-dimensional prismatic and tapered wall machinable features. When *MfgView* is selected, CAMWorks uses an alternative method to generate features and finds additional feature types not found by *AFR*.

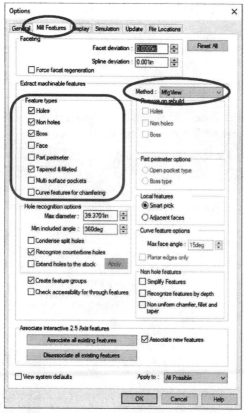

Figure A.2 The *Options* dialog box

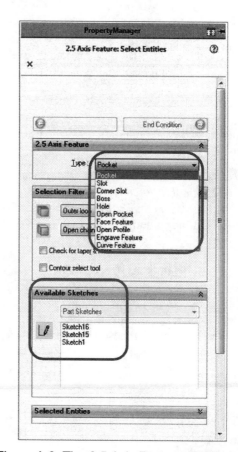

Figure A.3 The *2.5 Axis Features* dialog box

The following feature types are currently supported in CAMWorks:

- Bosses and pockets with vertical walls.
- Bosses and pockets with constant tapered walls with or without constant radius bottom or top fillets and chamfers.
- Pockets and bosses are further broken down into rectangular, circular, obround and irregular.
- Slots with vertical walls.

- Slots have categories for rectangular, corner rectangular, irregular, and corner irregular.
- Numerous hole types including simple holes, counterbores, countersinks, and multi-stepped holes.
- Simple holes can also be described as being drilled, bored, reamed, or threaded.

Local Feature Recognition (LFR)

CAMWorks provides a selective form of Automatic Feature Recognition (AFR) based on user-selected faces in the form of Local Feature Recognition (LFR). This is a semi-automatic method to define features based on face selection. Single or multiple features can be extracted depending on the selected faces.

You may pick one or more faces of the part on which local features are to be recognized. Then, right-click *Mill Part Setup* under the CAMWorks feature tree tab ▣ and select *Recognize Local Features* from the context menu. After executing this command, the locally recognized features, if any, will be added to the machinable feature list.

Interactive Feature Recognition (IFR)

Interactive Feature Recognition allows you to extract interactively either 2.5 axis or multi surface features.

As you may be aware, AFR cannot recognize every feature on complex parts and does not recognize some types of features. To machine these areas, you need to define features interactively by right clicking *Mill Part Setup* (under CAMWorks feature tree tab ▣) and choosing, for example, *New 2.5 Axis Feature*. In the *2.5 Axis Features* dialog box (Figure A.3), select feature type and pick a sketch of the feature to create a machinable feature interactively. Detailed steps for creating such a machinable feature can be found in the book, for example, Lesson 2: Simple Plate.

If you activate CAMWorks *3 Axis Milling* (as discussed in Lesson 1, see Figure 1.14), you may define multi surface features interactively by specifying faces to be cut and faces to avoid. Detailed steps for creating such machinable and avoid features can be found in Lesson 4: Machining a Freeform Surface.

As discussed in Lesson 1, CAMWorks uses a set of knowledge-based rules to assign machining operations to machinable features. The technology database (or TechDB™) contains data and rules that determine operation plans for the respective machinable features. These data and rules can be customized to meet your specific needs. You may refer to Lesson 10 for more details. The allowable milling operations for the respective machinable features established in TechDB™ are summarized in Table A.1.

Table A.1 A list of allowable milling operations for the respective machinable features

Machinable Feature	Description	Allowable Operations
Rectangular Pocket	A pocket whose general shape is a rectangle. The corners of the rectangle can be either sharp or radiused.	Rough Mill, Contour Mill
Circular Pocket	A pocket whose shape is defined by a circle. The difference between a circular pocket and a hole is that a circular pocket can contain an island.	Rough Mill, Contour Mill, Drill, Bore, Ream, Tap, Center Drill, Countersink
Irregular Pocket	A pocket whose shape is neither rectangular nor circular.	Rough Mill, Contour Mill
Obround Pocket	A rectangular shaped boss with 180 degree round ends.	Rough Mill, Contour Mill

Table A.1 A list of allowable milling operations for the respective machinable features (cont'd)

Machinable Feature	Description	Allowable Operations
Rectangular Slot	A pocket with one edge open to the outside of the part. Machining is extended beyond the open segment of the slot. The general shape of the feature is rectangular. The corners of the rectangle can be either sharp or radiused.	Rough Mill, Contour Mill
Irregular Slot	Similar in definition to a Rectangular Slot except that the general shape is not rectangular.	Rough Mill, Contour Mill
Rectangular Corner Slot	Similar to a Slot except that the feature can contain two or more adjacent edges that are open to the outside of the part. Machining is extended beyond each of these open edges. The corners of the rectangle can be either sharp or radiused.	Rough Mill, Contour Mill
Irregular Corner Slot	Similar to a Rectangular Corner Slot except that the general shape is not rectangular.	Rough Mill, Contour Mill
Rectangular Boss	A Boss whose general shape is rectangular. The corners of the rectangle can be either sharp or radiused.	Contour Mill
Circular Boss	A Boss whose shape is round.	Contour Mill, Thread Mill
Irregular Boss	Any Boss whose shape is neither rectangular nor circular.	Contour Mill
Obround Boss	A rectangular shaped boss with 180 degree round ends.	Contour Mill
Hole	Same shape as a SOLIDWORKS Solids Simple hole with a blind or through end condition or a Simple Drilled hole. Strategies can be assigned as Drill, Bore, Ream, or Thread.	Rough Mill, Contour Mill, Drill, Bore, Ream, Tap, Center Drill, Countersink, Thread Mill
Countersunk Hole	Same shape as a SOLIDWORKS Solids Countersunk hole with a blind or through end condition or a C-Sunk Drilled hole. Strategies can be assigned as Drill, Bore, or Ream.	Rough Mill, Contour Mill, Drill, Bore, Ream, Tap, Center Drill, Countersink, Thread Mill
Counterbored Hole	Same shape as a SOLIDWORKS Solids Counterbored hole with a blind or through end condition or a C-Bored Drilled hole. Strategies can be assigned as Drill, Bore, or Ream.	Rough Mill, Contour Mill, Drill, Bore, Ream, Tap, Center Drill, Countersink, Thread Mill
Countersink/Counterbore Combination	Each countersink/counterbore combination will be recognized as 2 separate hole features that are based on the Mill Part Setup direction.	Rough Mill, Contour Mill, Drill, Bore, Ream, Tap, Center Drill, Countersink, Thread Mill
Multi-stepped Hole	Any hole with multiple steps that is not recognized as a hole, countersunk hole or counterbored hole.	Rough Mill, Contour Mill, Drill, Bore, Ream, Tap, Center Drill, Countersink, Thread Mill
Open Pocket	Created only when AFR finds a Boss feature and the machining direction is parallel to one side of the stock. The bottom of the open pocket is the bottom of the boss. The boss becomes an island in the open pocket.	Rough Mill, Contour Mill

Table A.1 A list of allowable milling operations for the respective machinable features (cont'd)

Machinable Feature	Description	Allowable Operations
Face Feature	If the Face option is checked on the Features tab in the Options dialog box, a Face Feature is created when AFR finds at least one non-face machinable feature, the topmost face is parallel to the Mill Part Setup and the machining direction is parallel to one of the sides of the stock.	Face Mill, Rough Mill, Contour Mill
Perimeter–Open Pocket or Perimeter-Boss Feature	If the Perimeter option is checked on the Features tab in the Options dialog box, AFR creates either a boss or open pocket feature for the perimeter for any Setup created via AFR.	Rough Mill, Contour Mill, Face Mill

A.3   Machinable Features for Turning Operations

Similar to milling operations, machinable features can be extracted in SOLIDWORKS CAM either automatically using Automatic Feature Recognition (AFR) or created interactively by defining a turn feature from a SOLIDWORKS part face, sketch or edge.

The following turn feature types are currently supported:

- Face Feature: A Face feature is defined from vertical edges at the front edge of the part model.
- OD Feature: An OD feature includes the outside shape of the part from the face feature to the cutoff feature, not including the shape of any groove features.
- ID Feature: An ID feature includes the inside diameter shape of the part from the Face feature to the Cut Off feature, not including the shape of any groove features.
- Groove Feature with vertical walls: A Groove is a feature that is closed on both sides and below the surface of the surrounding geometry. There are three categories of grooves: rectangular, half obround and generic. You can cut a groove into the outer diameter, the inner diameter or a face. Groove features where the bottom of the groove is not parallel to the Z or X axis are not recognized by AFR and must be defined interactively.
- Cut Off Feature: A Cut Off feature is defined from vertical edges on the opposite side of the Face feature. A Cut Off feature is similar to a face and can be converted to a Face feature for two-step turning operations using the Convert to Face Feature command on the Cut Off feature context menu.

Some of the aforementioned features are illustrated in Figure A.4. The allowable turning operations for the respective machinable features established in TechDB™ are summarized in Table A.2.

Table A.2 A list of allowable turning operations for the respective machinable features

Machinable Feature	Description	Allowable Operations
Face	A Face feature is defined from vertical edges at the front of the part model. This feature is generally used to trim excess stock off the front of the part.	Face Rough Face Finish
OD	An OD (outer diameter) feature includes the outside shape of the part from the face feature to the cutoff feature, excluding the shape of any groove features. Typically, this feature is used to trim the stock away, leaving you with the outer shape of your design.	Turn Rough Turn Finish Face Rough Face Finish Groove Rough Groove Finish, Thread

Table A.2 A list of allowable turning operations for the respective machinable features (cont'd)

Machinable Feature	Description	Allowable Operations
ID	An ID (inner diameter) feature includes the inside diameter shape of the part from the face feature to the cutoff feature, excluding the shape of any groove features.	Bore Rough Bore Finish Drill Thread
Groove Rectangular OD	A Groove Rectangular OD is a recessed feature on the OD of the part where the side walls are equal and parallel to the X machining direction. The bottom and or top may include constant radius fillets or equal size chamfers.	Groove Rough Groove Finish Turn Rough Turn Finish
Groove Rectangular ID	A Groove Rectangular ID is a recessed feature on the ID of the part where the side walls are equal and parallel to the X machining direction. The bottom and or top may include constant radius fillets or equal size chamfers.	Groove Rough Groove Finish Bore Rough Bore Finish
Groove Rectangular Face	A Groove Rectangular Face is a recessed feature on the front edge of the part where the side walls are equal and parallel to the Z machining direction. The bottom and or top may include constant radius fillets or equal size chamfers.	Groove Rough Groove Finish
Groove Half Obround OD	A Groove Half Obround OD is a recessed feature on the OD of the part where the side walls are equal and parallel to the X machining direction. The side walls of the groove are joined at the bottom of the groove by a single 180 degree radius.	Groove Rough Groove Finish
Groove Half Obround ID	A Groove Half Obround ID is a recessed feature on the ID of the part where the side walls are equal and parallel to the X machining direction. The side walls of the groove are joined at the bottom of the groove by a single 180 degree radius.	Groove Rough Groove Finish
Groove Half Obround Face	A Groove Half Obround Face is a recessed feature on the Face of the part where the side walls are equal and parallel to the Z machining direction. The side walls of the groove are joined at the bottom of the groove by a single 180 degree radius.	Groove Rough Groove Finish
Groove Generic OD	A Groove Generic OD is any recessed groove shaped feature on the OD of the part that is neither rectangular nor half obround in shape.	Groove Rough Groove Finish Turn Rough Turn Finish
Groove Generic ID	A Groove Generic ID is any recessed groove shaped feature on the ID of the part that is neither rectangular nor half obround in shape.	Groove Rough Groove Finish Bore Rough Bore Finish
Groove Generic Face	A Groove Generic Face is any recessed groove shaped feature on the face of the part that is neither rectangular nor half obround in shape.	Groove Rough Groove Finish
Cut Off	A Cut Off feature is defined from vertical edges on the opposite side of a Face feature. This feature is used primarily to trim excess stock from the back of the part. A Cut Off feature is similar to a face and can be converted to a Face feature for two-step turning operations using the Convert to Face Feature command on the Cut Off feature context menu.	Cut Off Face Rough Face Finish Groove Rough Groove Finish

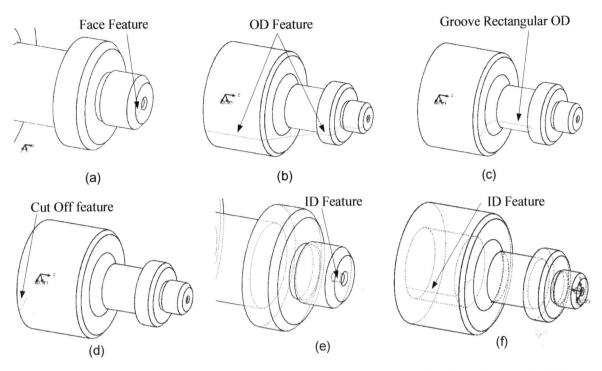

Figure A.4 Examples of machinable features for turning operations, (a) *Face Feature*, (b) *OD Feature*, (c) *Groove Rectangular OD*, (d) *Cut Off Feature*, (e) *ID Feature* (front end hole), and (f) *ID Feature* (rear end hole)

Notes:

Appendix B: Machining Operations

B.1 Overview

When a machinable feature is extracted or created, the corresponding machining operations are generated by the knowledge rules and data stored in the TechDB™. In this appendix, we discuss briefly the operations involved in both milling and turning operations.

B.2 Milling Operations

Milling operations are grouped by the types of machinable features, i.e., 2.5 axis features, 3 axis cutting cycles, and multiaxis.

2.5 Axis Features

The operations that support machining 2.5 axis features include Rough Mill, Contour Mill, Face Mill, Thread Mill, and Single Point.

A Rough Mill operation removes material from a part by following the shape of the machinable feature (pocket-in, pocket-out, spiral in or spiral out) or by making a series of parallel cuts across the machinable feature (zigzag, zig or plunge rough).

A Contour Mill operation removes material from a part by following the shape of the profile of pockets, slots, bosses, etc.

The Face Mill operation generates a toolpath on a mill Face feature for squaring or facing off the top of a part. Although a Rough Mill operation can generate toolpaths that will remove material from a Face feature, the Face Mill operation provides specific controls to produce a toolpath motion that is more appropriate for this task.

Thread Mill operations create thread mill toolpaths for a hole or circular boss. A thread mill operation can be generated automatically by assigning a thread mill Strategy to a hole or circular boss and selecting Generate Operation Plan. Alternatively, you can insert a thread mill operation using the New Operation and New Hole Operation commands.

Single point cycles include Drill, Bore, Ream, Tap, Countersink, and Centerdrill.

3-Axis Cutting Cycles

The 3 Axis cutting cycles include Area Clearance, Pattern Project, Z Level, Constant Stepover, Pencil Mill, Curve Project, and Flat Area.

The Area Clearance cycle removes the material between the stock or contain area and the selected feature at decreasing Z depth levels by making a series of parallel cuts across the stock (Lace) or by pocketing out toward the stock.

The Pattern Project operation is a multi-surface finishing cycle that removes material based on the selected pattern: Slice, Flowline, Radial and Spiral. These patterns have unique characteristics that make them appropriate for semi-finishing and finishing selected areas or the entire model.

The Z Level cycle is a finish contouring cycle that removes material by making a series of horizontal planar cuts. The cuts follow the contour of the feature at decreasing Z levels based on the Surface Finish you specify. Cutting starts from the highest location on the model and works down.

The Constant Stepover operation removes material by maintaining a constant user-defined stepover relative to the surface.

The Pencil Mill cycle generates toolpaths to finish machine corner areas using a single pass or multiple constant stepover passes. Corner areas are defined where the radius of curvature on the feature is less than the radius of the tool.

The Curve Project cycle removes material by projecting selected 2.5 Axis Engrave and/or Curve features onto the faces/surfaces of a Multi Surface feature and generating toolpaths along the projected entities. CAMWorks can calculate a single pass or multiple depth passes for engraving.

The Flat Area cycle uses a pocket out pattern to remove material from feature faces that are flat and parallel to the XY machining plane. CAMWorks generates toolpaths only on completely flat areas. If a face/surface has even a small gradient, CAMWorks will not generate a toolpath. This cycle can be used for finishing where excess material has already been cleared and supports single or multiple depths of cut.

Multiaxis Operations

Multiaxis operations are generated to machine a single freeform or multiple surface features to achieve desirable surface finish.

B.3 Turning Operations

The Turn cutting cycles include Face Rough, Face Finish, Turn Rough, Turn Finish, Groove Rough, Groove Finish, Bore Rough, Bore Finish, Drill, Center Drill, Thread, and Cut Off.

Face Rough: The Face Rough Operation defines multiple rough cuts to machine turned faces.

Face Finish: The Face Finish Operation defines a single finish cut to machine turned faces.

Turn Rough: The Turn Rough operation defines multiple rough cuts to machine turned faces. This is a common machining operation for a part that has a range of stock on the ODs that requires several cuts to remove the bulk of the stock.

Turn Finish: The Turn Finish operation defines multiple rough cuts to machine turned faces.

Groove Rough/Finish: The Grooving Cycle (Rough and Finish) allows you to cut a groove of almost any shape.

Bore Rough/Finish: The Bore Cycle (Rough and Finish) allows you to bore a hole or an ID feature.

Drill: The Drill operation generates a drill toolpath at the center line of the part (Z axis).

Center Drill: The Center Drill operation generates a center drill toolpath at the center point of the hole with a small depth.

Thread: In CAMWorks Turning, you use Strategies to define a thread on a feature. In order to generate a Thread operation, the corresponding thread condition must be selected from the TechDB™. The major diameter for the thread condition in the TechDB™ must match the feature's maximum diameter.

Cut off: When the stock is defined as bar stock, Automatic Feature Recognition generates a Cut off feature on the opposite side of the Face feature. A Cut off operation is generated for a Cut off feature when you generate an Operation Plan.

Notes: